Constructive Interference
Developing the brain's telepathic potential

Mark Fox BSc (Hons) First Class

Paperback edition first published in the United Kingdom in 2019
by aSys Publishing

Hardback edition first published in the United Kingdom in 2019 by
aSys Publishing

eBook edition first published in the United Kingdom
in 2019 by aSys Publishing

A CIP catalogue record for this book is available from
the British Library.

ISBN: 978-1-913438-04-3

aSys Publishing 2019

http://www.asys-publishing.co.uk

Favour rewarding; use punishment sparingly,
and only to define the limits.

Contents

Preface   vi

Part I: **Fabric**   1
1   Pregalactic medium   2
2   Electrical phenomena   12
3   Configurations   22

Part II: **Blind Spots**   33
4   Fissile material   34
5   Reforming chaos   44
6   Hating the light   54

Part III: **Neocentrism**   65
7   Separation, fragmentation and denial   66
8   Political middle-ground   77
9   Freedom of expression   86

Part IV: **Open-Competition**   97
10   Conceptual states   98
11   Progressive structures   106
12   Battle of the sexes   115

Part V: **Emotional Content**   125
13   Cogito, ergo sum   126
14   Sensory deprivation   134
15   Basal responses   143

Part VI: **Genome**   155
16   Dominant and recessive   156
17   Women's best instincts   166
18   A 'godly' conclusion   176

Appendix   185
Notes   252

Preface

As students, laypersons and practitioners all benefit from standard texts containing all the essential concepts, this book conscientiously mirrors the same. Accordingly, this extrasensory curriculum begins with the *dark sciences*, replete with their pioneering atomic model and theory of everything. Then, utilizing just four astrophysical fields and four sub-atomic particles, the dark sciences skillfully explain how life began as a trial-and-error synergy between nucleic acids and proteins. Central to that synergy were electric and magnetic fields – fields which remained resolutely 'on-top' throughout, sufficient to explain the subsequent evolution of human consciousness. Indeed, the dominant nature of those fields is a recurring theme throughout this book, with a latent telepathic potential seen to predictably arise.

As regards personality and behaviour, they derive from the neurophysiological conjuncture of three major communication systems – hence the book's timely analysis of *radiotransmission*, *neurotransmission* and *arteriotransmission* (that conjuncture clinically justifying the book's coining of the term *cognitive-nervous-endocrine axis*). Such is the commanding nature of the electric and magnetic fields, that remote manipulation, mental telepathy and the superimposition of persons are all technically achievable. However, while your body self-regulates around physiological norms, your mind is susceptible to mental illness due to the *healthy mind paradox*. That all-too-human paradox forcing a wholesale revision of these *radiofrequency effects*, as society's command of extrasensory phenomena brings human intelligence increasingly into question.

PART I

Fabric

Chapters 1-3

Ch1: **Pregalactic medium**

- Oscillating fields

The history of **science**, both theoretical and applied, runs parallel with philosophy's intense scrutiny of human reasoning. Accordingly, while **scientists** explore the substance of reality, frequently using the language of mathematics, **analytic philosophers** examine the associated world of statements, propositions and arguments, looking for logical inconsistencies. However, science and philosophy have a dilemma – namely, how to apply their knowledge, so painstakingly arrived at, within the wider socioeconomic environment. One answer, is to allow ill-judged intuition, rival faiths and opposing ethical axioms to determine how such knowledge is utilized – and all under the rather questionable pretext of democracy. However, given the alarming nature of contemporary technical know-how, many might question that rather cavalier approach.

Of course, a reasoned stratagem would make the future more predictable, the world considerably safer and human progress all but assured. But, in order to arrive at such a solution we must first objectively critique the current scientific paradigm, reformulate some of its theories and abandon several of its outdated nostrums. From its classical obsession with **illusory correlations** to habitually extrapolating from fragmentary experimental data, orthodox science has unwittingly shrouded the truth in popular misconceptions. Only by discarding those fallacies can we regroup around irrefutable facts, before further developing scientific expertise, but in an altogether more intelligible manner.

Central to our quest for defining answers, in an age of growing telepathic awareness, is the pursuit of unorthodox conjectural models – ones which accurately reflect the true realities. Our universe, from its very inception, was marked by spontaneous decay, but not wholesale disintegration – the fabric of **space** quickly arose, with **atoms** denoting a new-found stability. It's reasonable to infer that only a finite number of fields exist and that a preordained **symmetry** forces atoms into existence. Today, the terms **mass** and **energy** are no longer sufficient to fully explain the dynamics of time, space and matter. What's missing, of course, is that other telling feature of the cosmos – namely, its appetite for expansion. Account for that, and a far more insightful picture emerges.

The term **medium** has many connotations, ranging from a state between extremes or a material through which energy is conveyed, to a means of preserving the past or bringing about specific outcomes. Our universe embodies all of those attributes, making it, without doubt, the most fascinating medium of all. The **pregalactic medium** – that intriguing precursor to today's **galactic** and **intergalactic** mediums – proved to be a blank canvas upon which galaxies could easily form (the subsequent engine of stellar formation and death enabling highly-concentrated fields, forces and particles to spontaneously differentiate into atoms of varying **elements**). That chemical enrichment being contemporaneous with energy's transmutation into the **electromagnetic spectrum**.

Contrary to convention, one argues that **protons** mutually attract, creating mass, and that **electrons** mutually repel, creating **solidity**.[1] And so, whilst protons accelerate a painful fall, it's actually the electrons which bring you embarrassingly to a halt. Accordingly, **fission** doesn't convert mass into energy, it simply expresses mass outwardly, and in a particularly energetic fashion – the protons violently associating with other, more distant, protons. That's not to imply that mass never transmutes into energy, or *vice versa*, it's just that it could only ever happen at the astrophysical extremes, either deep within stars or unseen within the vacuum of space. That is to say, by working from the general to the particular, this narrative arrives pioneeringly at the **dark sciences**.

- Natural curiosity

In the late 1920s, Edwin Hubble (1889-1953), an American astronomer based at the Mount Wilson Observatory in Los Angeles, pointed what was then the world's most powerful telescope at the night sky, and discovered that stars were concentrated within discrete galaxies. The light emitted by several extraordinarily distant galaxies was then studied in some detail, sufficient to establish that the **wavelengths** were becoming stretched-out or elongated. That elongation, or 'red-shift', ineluctably proving that these galaxies were moving away from the earth, and at increasing velocities as one peered ever deeper into the night sky. This so-called Hubble expansion suggested that all discernable energy and mass was once concentrated within a much smaller space. Our best estimate as to when this '**big bang**' expansion first began is 13.7 billion years ago (this being the projected age of our universe).

Around the time of Hubble's ground-breaking discovery, the world witnessed the birth of modern **chemistry**. Negatively-charged electrons, positively-charged protons and electrically-neutral **neutrons** were discovered, enabling the principles governing their sub-atomic interactions to be ingeniously worked-out. Thus, while the public reflected on the ramifications of a 'big bang' – an idea first proposed by Georges Lemaitre (1894-1966), in which he suggested that mass and energy were once concentrated within some primordial 'cosmic egg', prior to its explosive decay – science began to focus on how sub-atomic particles might actually have formed. The thermal remnants of that blast becoming known as the **cosmic background radiation**, with mounting evidence for the same appearing to vindicate increasingly costly attempts at replicating 'big bang' effects under laboratory conditions.

Convinced that the secrets of matter lay hidden within those highly-energetic conditions, **particle acceleration** became the vogue from the 1930s onwards. Armed with a late-Victorian grasp of **electromagnetism**, interwar verification of cosmic expansion and an ever-growing list of **sub-atomic particles**, scientists began proposing ever more sophisticated methods for creating, accelerating and directing sub-atomic particles. The first accelerators, e.g. cyclotron, voltage multiplier and electrostatic generator, had energies of just a few million **electron volts** (MeV), but were soon eclipsed by more powerful accelerators, generating billions of electron volts (GeV). Predictably, come the 1980s, a 'Superconducting Supercollider' was planned for Texas, which would've been humanity's most costly scientific enterprise to date. In the event, US Congress halted its construction, whereupon the initiative passed to those scientists working on Europe's Large Hadron Collider (LHC).

The Large Hadron Collider (a 27km ring-shaped particle accelerator, sited at CERN, in Geneva) is capable of generating energies in the order of tens of trillions of electron volts (TeV). Whether the field of analytic philosophy is capable of keeping-up with these developments and objectively scrutinizing all of the associated statements, propositions, arguments and algebra remains an open question. For example, in 2012, a particle, called the Higgs boson, was reportedly discovered by scientists working at the LHC. This particle, and the associated Higgs field, were said to interact with conventional particles, affording them mass. The dark sciences would've saved them an enormous amount of money, by arguing that the mutual attraction of protons is contingent upon known fields.

4

- Analytic philosophy

Philosophy is more concerned with the defining attributes of truth than science. For example, **peer review** is perceived as foolproof by science, whereas philosophy is decidedly more **sceptical**. However, where science and philosophy diverge the most, is over the question of whether mathematics can successfully replicate reality. But if scientific peers aren't analytic philosophers, what are they? Scientific peers are, in fact, experts in a particular field who are distinguished enough to be asked by editors to review academic papers. Thus, editorial interference remains a weakness, as does favouritism towards well-known academics and researchers. These criticisms are important, as they help us to understand how paradigms first arise – and, more importantly, how they quickly unravel.

Even in the absence of editorial interference or favouritism, science still risks arriving at **working hypotheses** which aren't strictly falsifiable. There was, for example, a time in our historical past when the prevailing **hypothesis** that "*the sun goes around the earth*" wasn't practicably refutable. Had emergent science tested that particular proposition, it may have found weak evidence in support of the same. At which point, having failed to disprove that particular assertion, it would then have been moving cautiously towards an as-yet-unproven nonsense. This is the great difficulty of extrapolating from experimental data, in-so-far as '*big bang creationism*' is difficult to refute, but was it ever properly refutable in the first place? If not, it may prove, like **supersymmetry**, to be a compelling nonsense.

The peer review process can be profitably appraised using both a **rationalist** and **empiricist** perspective. Crudely put, **empiricism** sees knowledge as drawn from direct experience, via one's senses, whereas **rationalism** sees knowledge as a largely non-sensory product of reason. Accordingly, rationalists take an *a priori* approach to the truth, using language, mathematics and logic to explore and understand the same. Conversely, empiricists adopt an *a posteriori* position, accepting as true only such things as they can, or could, personally sense. Arguably, the peer review process is rationalist, being heavily reliant upon language, mathematics and logic – together with that poorly-understood 'faster-than-light' phenomenon termed trust.

In truth, we find that actual scientific research straddles the fine line separating rationalism and empiricism, in-so-far as an *a priori* working hypothesis (arrived at through the shrewd use of language, logic and reason) is then tested by experiment. Those experimental observations are largely sensory, or else involve technology's rapidly expanding list of sensors, sufficient to test the researcher's capacity for *a priori* conjectural thinking. In other words, the **scientific method** is, to all intent and purposes, a test of our ability to anticipate the truth – which both explains and justifies the adoption of falsifiable hypotheses consistent with one's expectations.

Thus, knowledge is what happens when the rationalist and empiricist collaborate, with **double-blind controls** guaranteeing impartial analysis. Nevertheless, the 'big bang creationists' are right in one vital regard – understand the origins of mass, energy and expansion, and one can arrive at countless working hypotheses, many of which would survive **experimental testing**. Accordingly, this book rethinks the origins of those media providing for **remote manipulation**, **mental telepathy** and the **superimposition of persons**. Or, should I say, it constructively fuses rationalist and empiricist thinking, sufficient to arrive at an all-encompassing **theory of everything**. A theory which not only accounts for **telepathically-induced effects**, but which also champions their enlightened application.

•   Theory of everything

During **fusion**, surplus mass radiates-out as **light**. According to the dark sciences, surplus light can, in fact, precipitate-out as **mass**. For this to happen, mass *must* contain all the constituents of light, and light *must* contain all the ingredients of mass. Consequently, at the astronomical extremes of attraction and repulsion, mass and energy become interchangeable. **Gravity** (the gravitational field in its stronger galactic form) supports and produces light, whilst concentrating mass. **Gravitation** (the gravitational field in its weaker intergalactic form) forces light's conversion into mass, amid burgeoning forces of repulsion. The **galactic medium** is therefore dominated by the mutual attraction of protons, and the **intergalactic medium** by the mutual repulsion of electrons (respectively termed **gravitational forces** and **dispersional forces**).

The equation $E=mc^2$ (or $m=E/c^2$) reflects the fact that energy and **mass-dispersion** (gravitational-dispersional forces) are equivalent, but by no means equal. The **dark cycle theory of everything** (see Fig. 1) proposes that energy **transduces**, in the presence of gravitation, into mass-dispersion (or, more specifically, that electromagnetic waves transmute into atoms when unsupported by gravity, with accompanying inflationary effects). Therefore, in the absence of gravity, oscillating electric and magnetic fields spontaneously break-down into charged particles – the **electric field** creating dispersive electrons and the **magnetic field** creating gravitating protons (with any neutrons produced being susceptible to **beta decay**, whereby protons and electrons are ejected). Central to this **deep-space nucleosynthesis** is the fabrication of space itself – a medium which comprises forces of both contraction and inflation.

Figure 1: The Dark Cycle: theory of everything

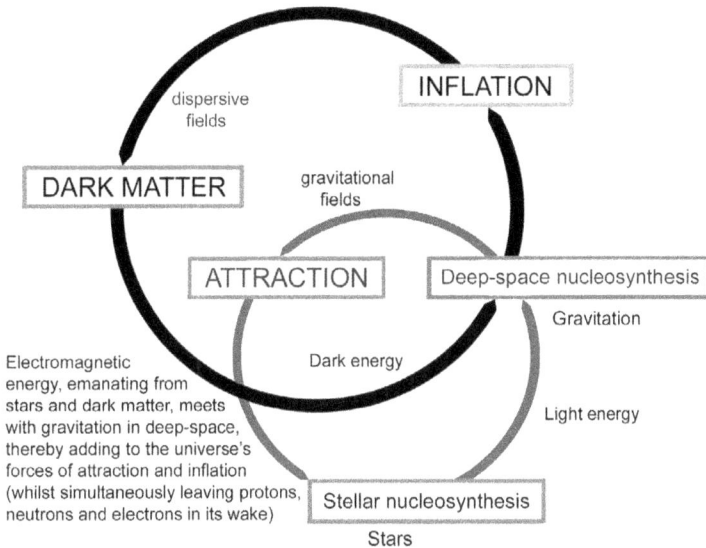

dispersive fields

INFLATION

DARK MATTER

gravitational fields

ATTRACTION

Deep-space nucleosynthesis

Gravitation

Electromagnetic energy, emanating from stars and dark matter, meets with gravitation in deep-space, thereby adding to the universe's forces of attraction and inflation (whilst simultaneously leaving protons, neutrons and electrons in its wake)

Dark energy

Light energy

Stellar nucleosynthesis

Stars

Wave-particle duality implies that sub-atomic particles possess wave-like characteristics and that light possesses the characteristics of so many particles. This allows us to think of light – or, more technically, **electromagnetic radiation** – as comprising **photons**, **transverse waves** or **oscillating** electric and magnetic fields. Those fields, waves and particles convey specific units of energy. In fact, everything one needs to fabricate sub-atomic particles in deep-space. **Electrostatic attraction**, together with the stabilization of neutrons within simple nuclei, explains the spontaneous generation of hydrogen and helium atoms. What today's astronomers would term interstellar gas – interstellar gas rivalling the best vacuums produced here on earth, whilst being nonetheless sufficient to seed new galaxies. Central to that galactical accretion are atoms, whose configuration does much to mediate the course of events. If there are too few atoms, as characterized by the intergalactic medium, light will spontaneously transform into electrons, protons and neutrons. If atoms are abundant, as characterized by the galactic medium, light waves will flourish – **wave intensity** denoting the amount of energy carried by those waves, such energy reflecting both their **frequency** and **amplitude**. Energy's conversion into mass-dispersion renders the light-source invisible, hence the phenomenon known as **dark matter**. **Dark energy** is simply the wave-intensity driving that deep-space nucleosynthesis. In essence, light arriving here on earth has passed through an astrophysical filter, leaving expansion in its wake.

Generally-speaking, chemical **bond formation** releases energy, whilst chemical **bond-cleavage** requires energy. At the astronomical scale, **stellar nucleosynthesis** expels energy, whilst deep-space nucleosynthesis locks energy away. Accordingly, the universe possesses an underlying symmetry, or **conservation**, by virtue of mass inhibiting mass production, through light sustaining light's propagation, and owing to the sum of its attraction equalling the sum of its repulsion. However inflationary its empirical appearance, energy's transmutation into mass-dispersion guarantees a comparable amount of contraction. Which implies that the universe's ability to generate energy isn't easily diminished. Who knows, perhaps gravitational collapse will one day bring renewed order and available energy in place of unalleviated disorder and **heat death**.

- ## Substance of reality

Energy is a relatively hollow concept without some form of medium to provide for its transmission, conversion and utilization. Thus, gravitational potential energy, kinetic energy, electrical energy, chemical energy and thermal energy all rely on moving objects, flowing electrons, chemical bonds and conductive materials. In many ways, energy avoided becoming a worthless abstraction due to the spontaneous proliferation of charged particles – particles which readily form into atoms, due to escalating electrostatic attraction in deep-space (the resultant hydrogen and helium gases experiencing variable forces of attraction and repulsion, amid fluctuating electric and magnetic fields).[2] Undisturbed, atoms and molecules balance-out these forces. However, the role of energy is to tirelessly disturb the same, often electrically (see Fig. 2).

Figure 2: Electricity: a cosmological construct

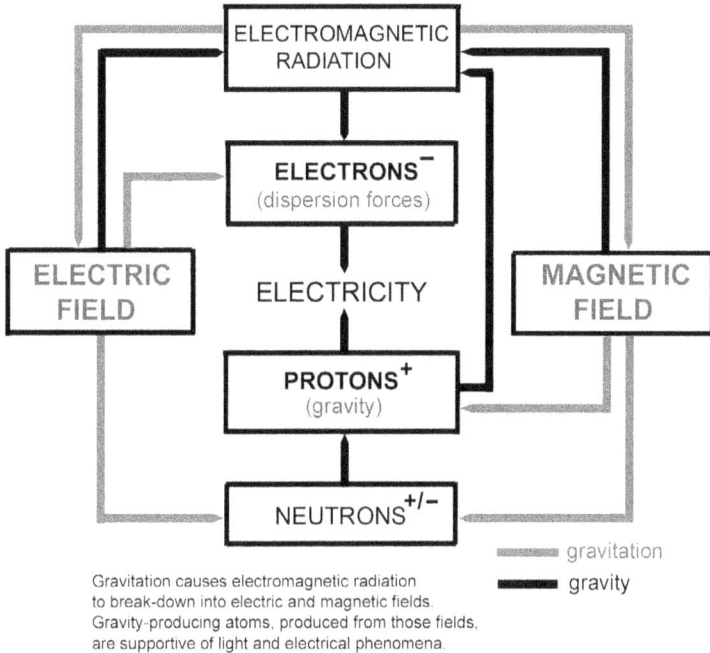

```
        ┌──────────────────────┐
        │  ELECTROMAGNETIC     │
        │     RADIATION        │
        └──────────────────────┘

              ┌──────────────┐
              │  ELECTRONS⁻  │
              │(dispersion forces)│
              └──────────────┘

┌─────────┐                        ┌─────────┐
│ ELECTRIC│      ELECTRICITY       │ MAGNETIC│
│  FIELD  │                        │  FIELD  │
└─────────┘                        └─────────┘

              ┌──────────────┐
              │  PROTONS⁺    │
              │  (gravity)   │
              └──────────────┘

        ┌──────────────────────┐
        │    NEUTRONS⁺/⁻        │
        └──────────────────────┘
```

▓▓▓▓▓ gravitation
■■■■■ gravity

Gravitation causes electromagnetic radiation
to break-down into electric and magnetic fields.
Gravity-producing atoms, produced from those fields,
are supportive of light and electrical phenomena.

The **strong nuclear force** is simply gravitational attraction at the sub-atomic scale (in other words, the point at which the value for 'distance' in the universal law of gravitation becomes zero, and the force of attraction between the **nucleons** increases astronomically). The strong nuclear force, **weight**, gravity and gravitation constitute a **spectrum of attraction**. Which is mirrored by a **spectrum of repulsion**, comprising the **weak nuclear force**, solidity, **dispersion forces** and **dispersive fields**. For example, weight is a characteristic of an object languishing on a celestial body, provided both possess protons and neutrons. What prevents them from collapsing together to a point is the mutual repulsion of their electrons. As for the weak nuclear force, that's associated with neutrons – those neutrons making a large atomic **nucleus** prone to to decay.

English-born scientist Sir Isaac Newton (1642-1727) and German-born physicist Albert Einstein (1879-1955) both grappled with the thorny question of how we satisfactorily explain gravitational attraction. Newtonian **physics**, for example, explains gravitational attraction in terms of a concise mathematical law, replete with gravitational constant (plus a unit of acceleration due to earth's gravity, termed Helmert's formula). Einsteinian physics, conversely, perceives gravitational attraction as resulting from distortions within the fabric of **space-time**, making gravitational forces geometrical rather than arithmetic. Taking the **dark cycle atomic model** as its starting point, this book now seeks to establish the true nature of astrophysical attraction and repulsion.

The atom's **electronic configuration** typifies the grey area separating physics and chemistry – chemistry being concerned with available energy, and hence fluctuating electric and magnetic fields, and physics being concerned with the constants, laws and invariants governing the particles present in those fields. Overlapping magnetic fields strengthen and weaken one another, producing a distinctive pattern of shells, subshells and **orbitals**. Consequently, the atom's electric fields, which respect those magnetic field lines, drive the electrons around biologically unique orbitals. Because **dispersional** and **gravitational** fields derive from electric and magnetic ones, the principles governing gravity and dispersion are much the same. That is to say, isolines and gradients compel sub-atomic particles in all manner of directions – often together, but frequently apart.

- Light source

The intergalactic genesis of **dispersional fields** is immediately inflationary. Conversely, **gravitational fields** take time to amalgamate and exert a discernable galactic influence. Whatever the scale, the electric and magnetic fields retain their primacy over nucleons and electrons – with nucleons and electrons exerting their influence via gravitational and dispersional forces, and through the profound impact of electrostatic attraction. **Galaxy clusters** (which bear testament to cosmological contraction equalling astronomical expansion) radiantly dispel one another. Deeply embedded within these galactic bodies, overlapping magnetic fields, arising in the course of organic processes, generate never-to-be-repeated **absorption profiles**, making them home to an array of cognizant life-forms. Needless to say, this book outlines how the telepathic potential, arising from those unique absorption profiles, can be responsibly harnessed.

10

To avoid the **paradoxical** pursuit of urgent answers by means of irresolute forms of reasoning, this book switches between rationalism and empiricism, whilst incorporating well-intentioned pragmatism. In other words, given the potential terrors of contemporary technical know-how, we mustn't be deflected from arriving at an enlightened model of social progress – one which can be sympathetically applied in the present, by **well-balanced** individuals. Analytic philosophy might argue that its purpose isn't to summon-up anthropological solutions, but merely to evaluate purportedly rational statements, looking for logical inconsistencies. If so, we must look to fill that particular vacuum ourselves – bearing in mind, of course, that the answer ought to satisfy the most scrupulous of logicians, the most exacting of scientists and the most anxious of parents.

Reassuringly, after scouring the arts, sciences and humanities, we do eventually arrive at a judicious answer, albeit half-hidden within the **human genome**. A genome which owes its existence to just four elementary particles (namely, electrons, protons, neutrons and photons) and just four fundamental fields (that is, the electric, magnetic, gravitational and dispersional). But to fully understand how those fields influence the said particles, sufficient to shape organic life, we must first add to the **laws of thermodynamics**, deconstruct the highly-deterministic electronic configuration and re-write the biography of energy-demanding chemical reactions. Should an absence of available energy make energy available, then the pregalactic medium will appear illusory, with **entropy** amounting to nothing more than a short-lived intellectual diversion.

Ch2: **Electrical phenomena**

• Unalleviated chemical enrichment

The **first law of thermodynamics** states that energy can neither be created nor destroyed (otherwise known as the law of **conservation of energy**). Similarly, there's a law of **conservation of mass**, which states that matter can neither be created nor destroyed. From the perspective of the dark sciences these laws pertain to chemical reactions, not fundamental physics. Of course, all the chemistry encountered here on earth conforms to these laws. However, such chemistry begins to break down once we meet with the weakest of intergalactic gravitation or the strongest of galactic gravity. The **second law of thermodynamics** states that the entropy of an **isolated system** will increase over time. This second law prohibits neutrons, once concentrated within low-energy helium nuclei, from spontaneously decaying. It also implies that the universe will one day become similarly inert, with a paucity of available energy.

As regards the **third law of thermodynamics**, that states that there's a minimum temperature at which "*the motion of the particles of matter would cease*". That minimum temperature would certainly put paid to chemistry, but would it materially impact upon gravitational and dispersional forces? The dark cycle **atomic model** proposes that electrostatic attraction increases as energy wanes. In other words, electrostatic attraction is inversely-proportional to energy, rather than distance – enabling the electric and magnetic fields to regulate how close an electron is to a given nucleus, and whether a neutron is formed. However, in order to fully comprehend that dynamic, we need disciplines which describe the impact of magnetic fields on nucleons and the impact of electric fields on electrons. Only then can we precisely model electrostatics and factor in the effects of gravitational and dispersional forces.

Quantum electrodynamics (QED) has traditionally sought to explain how light and matter interact. Given the revisions in this book, it is proposed that **QED** concerns itself with photon emission, electric fields, electrons and dispersional forces. **Quantum spatialdynamics** (QSD), on the other hand, would address photon absorption, magnetic fields, nucleons and gravitational forces. A photon is emitted whenever an atom's electric field weakens, sufficient for an electron to fall back into its **ground-state**. Conversely, an ambient magnetic field may strengthen an atom's ground-state magnetism, initiating the absorption of a photon – a case of incident radiation generating the conditions for its own absorption. **Absorption** and **emission** are therefore driven by a combination of strengthening magnetic fields and weakening electric fields.

QSD supersedes quantum gravity, in-so-far as it derives not from Einstein's general theory of relativity, but from the dark cycle theory of everything. Thus, quantum spatialdynamics examines how magnetic fields are able to transmute into gravitational forces, together with so many protons. Whilst quantum electrodynamics investigates how electric fields are able to transmute into dispersional forces, together with so many electrons. All assuming, of course, that gravitation initiates that conversion. Gravity acts as a bridge to light – with dispersional fields working in tandem with it to both propagate and propel emitted radiation. Together, the gravitational and dispersional fields interact as **spatialgravitism** (mirroring the magnetic and electric fields, which interact as electromagnetism).

Electrostatic attraction is an aspect of spatialgravitism, whereas **electromagnetic induction** is an aspect of electromagnetism. In other words, electromagnetism and spatialgravitism represent the interface between QSD and QED. Thus, the academic fields of QSD and QED must collaborate on questions pertaining to induction and electrostatics, and in relation to questions concerning escalating **atomic mass** (the significance of which lies in the decay wrought by the weak nuclear force). More broadly, the **periodic table** reflects a whole spectrum of gravitational and dispersional waves, emanating from each and every chemical element. Galaxies, in spite of this unalleviated chemical enrichment, can't augment their mass, thus limiting their ability to propel light. What they can do, of course, is seed new galaxies.

• Fourth law of thermodynamics

The writer's proposed law of **conservation of electronics** states: "*the electric and magnetic fields, acting upon charged particles which are in motion, must be of a prescribed intensity, otherwise the said particles will jump or fall into different orbitals, or else realign in the case of atomic nuclei*". Accordingly, when an atom's electric field weakens, the electron must fall into a lower-energy orbital in order to experience the stipulated field strength. This proffered **fourth law of thermodynamics** helps us to understand absorption, emission, chemical bonding, bond-cleavage and potential difference. As we've seen, there's a minimum temperature, below which charged particles will stop moving. However, this suggested '**4ᵗʰ Law**' pertains to particles which are in motion.

This striving for uniformity, amid ever-fluctuating fields, dictates a particle's position in space, together with its momentum. Moreover, it also explains **Planck's constant**, radiating energy having a frequency and amplitude which is reflective of those sub-atomic positions and changes in momentum. As you might imagine, the fourth law of thermodynamics proposes that there's an **electronic constant**, pertaining to electrons, which subjects those negatively-charged particles to an **electromotive force**, or **e.m.f**, whenever the relevant electric fields fluctuate. Likewise, there's a **nucleonic constant**, pertaining to atomic nuclei, which subjects those positively-charged nuclei to a **nucleomotive force**, or **n.m.f**, whenever the relevant magnetic fields fluctuate (see Fig. 3).

Figure 3: The dark cycle atomic model

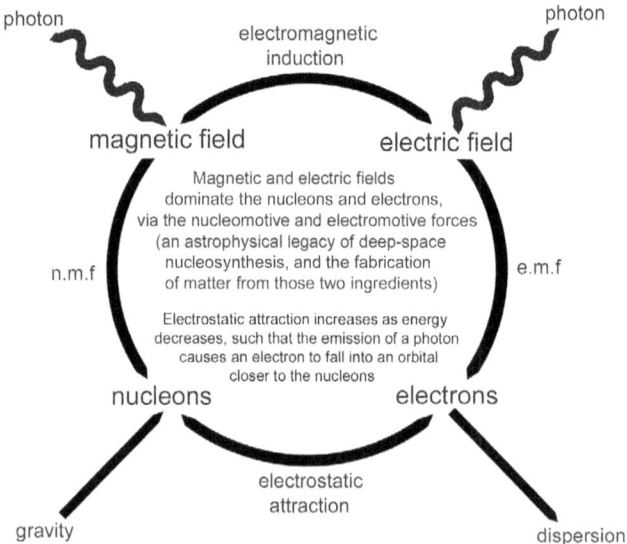

photon
photon
electromagnetic
induction
magnetic field
electric field

Magnetic and electric fields
dominate the nucleons and electrons,
via the nucleomotive and electromotive forces
(an astrophysical legacy of deep-space
nucleosynthesis, and the fabrication
of matter from those two ingredients)

n.m.f
e.m.f

Electrostatic attraction increases as energy
decreases, such that the emission of a photon
causes an electron to fall into an orbital
closer to the nucleons

nucleons
electrons

electrostatic
attraction

gravity
dispersion

Accordingly, incident radiation may strengthen an atom's magnetic field, initiating the absorption of a photon – whereupon electromagnetic induction will bolster the accompanying electric field, obliging one of the atom's electrons to jump-up into a higher-energy orbital (while the newly realigned nucleus reduces electrostatic attraction, facilitating the said jump). The atom has now achieved a more energized equilibrium. If, at some point, the electric field weakens, a photon will be emitted – whereupon electromagnetic induction will cause the depleted magnetic field to return the nucleus to its original alignment, increasing electrostatic attraction in the process, as the electron falls back into its original ground-state. Whereupon the atom has achieved a less energized equilibrium.

The electronic and nucleonic constants are unyielding, such that the electric and magnetic fields have enormous utility. For example, **magnetic resonance imaging**, or MRI, uses magnetic fields to realign a person's atomic nuclei, with the energy emitted, when those carbon and hydrogen nuclei 'relax', proving medically illuminating. But how does our *avant-garde* dark cycle atomic model explain something as mundane as a bar magnet? Well, thanks to the genius contribution of one's **heteronnubial** co-author, we can say that two bar magnets, with opposite or identical poles meeting, will attract or repel one another, because the fourth law of thermodynamics demands that their electrons move closer or further apart (of course, in non-magnetic materials the electromotive forces act in all directions and with much reduced intensity).

Ambient energy levels guarantee that electrons remain a discrete distance from atomic nuclei – the only proviso being that if energy levels are enormous the electrons must form a **plasma**, due to the sheer strength of the electromotive and nucleomotive forces (and due to a precipitous reduction in electrostatic attraction). In the case of electromagnetic induction, a varying magnetic field has the appearance of generating an electrical current in a circuit. In reality, the magnetic field only appears to have generated an electrical current. What the magnetic field has actually done is induce a **potential difference** in the electric field, which produces an electromotive force on the circuit's electrons. Bearing in mind, of course, that the magnetic field will also subject the current-carrying wire's nuclei to a nucleomotive force, loosening their hold over those electrons.

The '**zeroth law' of thermodynamics** states that if two thermodynamic systems are in thermal equilibrium with a third, they must also be in thermal equilibrium with one another. Which, in everyday terms, provides for thermodynamic systems being accurately measured. In truth, that '*conservation of calibration*' owes much to the electronic and nucleonic constants, which dictate how matter expresses temperature. However, those constants do far more than facilitating enumeration, they also bring breathtaking precision to biological processes. In **genetics**, for example, they enable **nucleic acids** to be transcribed and translated with remarkable consistency. Crucially, having evolved to exploit overlapping magnetic fields and variable electric fields, life has since rendered those fields ever more inimitable through time.

• Smashing particles

Chemical bonding, by which we mean **covalent bonding**, arises when electronegative atoms seek-out a lower-energy state, with the added bonus of stability (**electronegativity** denoting the strength with which an atom attracts electrons, sufficient to fill a partially-empty **valence shell**). Many of those electronegativity values increase from left to right as one views the periodic table of elements. This comparative **periodicity**, as regards electronegativity, suggests that chemical bonding is influenced by atomic mass, as atomic mass also increases from left to right (possibly reflecting complicated lines of magnetic influence in atoms of escalating **atomic weight**). What's clear, is that the peripheral magnetic fields weaken, forcing **valence electrons** into lower, more stable molecular orbitals, with the spontaneous emission of photons.

This makes chemical bond formation energy-releasing, with the electrons falling into **molecular bond-states**, rather than atomic ground-states. To subsequently cleave those bonds requires energy, by which we mean stronger and more powerful electromotive and nucleomotive forces. The periodic table of elements clearly possesses an intriguing substructure, what with gravitational and dispersional forces sabotaging otherwise straightforward groupings and periodicity, and with the magnetic manipulating the electric in an increasingly convoluted fashion. Of course, it's all very well talking about '*atomic weight*', but what of '*atomic solidity*'? That is to say, at least as important as atomic mass, is atomic repulsion. Account for that, and we'll understand the weak nuclear force, solidity, dispersion forces and dispersive fields (as they pertain to each and every chemical element and the **compounds** they form into).

Expanding upon that point, a projectile fired into soft clay penetrates so deep. If one doubles the projectile's velocity, it penetrates four times as deep. This relationship can be expressed as **$E=mv^2/2$**, i.e. the amount of **work** produced is proportional to the projectile's mass, times its velocity squared, divided by two. In this scenario, the projectile's kinetic energy is seen to make light work of the clay's dispersion forces – forces which would otherwise permit a stationary projectile to lie on a soft clay surface. However, before we run away with the notion that displacing electrons is easy, bear in mind that this scenario is less about kinetic energy, and more about the mutual repulsion of the electrons. In other words, we're actually smashing negatively-charged particles together, in order to better understand a whole spectrum of repulsion.

Therefore, mass-dispersion serves as an intriguing **independent variable**, one which confounds otherwise straightforward linear relationships. In fact, by taking electromagnetic energy and differentiating it into gravitational, kinetic, electrical, chemical and thermal alternatives, gravitational and dispersional fields are life-inducing antagonists, battling to conserve energy's hold over matter.

• Life's electronic blueprint

Taken together, the **Aufbau principle**, **Pauli exclusion principle** and **Hund's rule** stipulate: 1) that electrons occupy the lowest energy orbitals first, 2) that a given orbital only accommodates two electrons, and 3) that all orbitals, within a given shell or subshell, be filled by a solitary electron prior to the addition of a second. Given that those electrons are subject to electromotive forces, electrostatic attraction and dispersion forces, it's not difficult to see why the **azimuthal 'spdf' quantum number model** has become the *de facto* portrayal of sub-atomic orbitals (see Fig. 4). In truth, an atom's electric fields are strongly influenced by the atom's magnetic fields (which, in turn, are heavily influenced by the **magnetic field lines** of adjacent atoms). Accordingly, electronic configurations at the molecular level are biologically unique, hence humankind's latent telepathic potential.

Whatever an orbital's shape, its configuration isn't altogether flexible – but rather, sufficiently inflexible, so as to produce viscosity, ductility, malleability, elasticity, electricity and phases of matter. Indeed, so inflexible is the electronic configuration that if a current-carrying wire passes through a magnetic field, the wire will experience a force. By this means, an electric motor is able to generate mechanical energy, the motor's current-carrying coil having a reversible flow, which repeatedly interacts with a magnetic field, producing an uninterrupted mechanical force. At the heart of this device are unyielding sub-atomic particles, hence the significance of mutable isolines and gradients. Worryingly, that very inflexibility can also produce salinity, photochemical smog, acidification, podzolization, thermal pollution and mass extinctions.

Figure 4: Electronic configuration (and magnetic field lines)

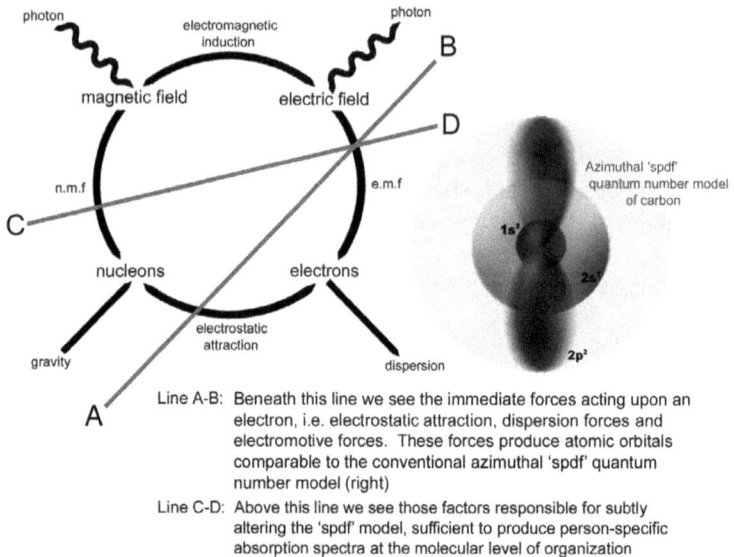

Line A-B: Beneath this line we see the immediate forces acting upon an electron, i.e. electrostatic attraction, dispersion forces and electromotive forces. These forces produce atomic orbitals comparable to the conventional azimuthal 'spdf' quantum number model (right)

Line C-D: Above this line we see those factors responsible for subtly altering the 'spdf' model, sufficient to produce person-specific absorption spectra at the molecular level of organization

In spite of these escalating anthropogenic concerns, organic chemistry has routinely harnessed scalars and vectors in order to inventively machinate at the molecular level. Thus, proliferating antagonistic forces, unalleviated chemical enrichment within the galactic medium and the merger of magnetic field lines within three-dimensional space have all augmented a much simpler electromagnetic spectrum – making life, as one's heteronnubial co-author intelligently posits, an offshoot of spectrally-enhanced electromagnetic induction. Basically, organic chemistry provides for the electric and magnetic fields vying with one another in an ever more animated manner. The ensuing sub-cellular reactions, multifarious organisms and codependent ecosystems flamboyantly expressing that underlying electromagnetic tension.

Gause's principle states that no two species can simultaneously occupy the same biological niche. Which implies that organisms within a **community** are **competitive**, rather than **competing**. In other words, a properly balanced ecosystem comprises competitive species, rather than competing ones. Nevertheless, individuals from the same, or closely-related species, may compete amongst themselves, provided it's to remain competitive. This suggests that competing humans, who disturb otherwise stable **ecosystems**, are ecologically uncompetitive. Thus, humans stress the planet through their impact on global temperatures (**climatic factors**), the soils (**edaphic factors**), marine environments (**oceanic factors**) and other living things (**biotic factors**). Making **anthropogenic factors** (human activity) increasingly worrisome.

What we find is that the elements from which **life cycles** are wrought – primarily oxygen, carbon, hydrogen, nitrogen and phosphate – are all part of much broader cycles. Consequently, the **hydrologic cycle**, **nitrogen cycle**, **carbon cycle** and **phosphorus cycle** all serve to bridge the gap between inanimate and animate, by maintaining the minimum conditions for life. Almost without exception, the energy driving these cycles comes from the sun, with its biological utilization becoming ever more sophisticated as one ascends through each successive **trophic level** – eventually terminating, at the very top, with the pathological complexities of **personality** and **behaviour** (personalities and behaviours with the propensity to compromise life-sustaining cycles, as happens when the nitrogen cycle is wantonly disrupted by the careless use of pesticides).

Without doubt, the earth's most competitive life-forms exploit solar energy optimally. What's more, they **adaptively-radiate**, filling all the available **ecological niches** in a sustainable manner. Viewed from an anthropological perspective, there are two types of person – those who maintain ecological stability and submit to being part of a competitive whole and those who compete rapaciously and threaten our environmental security. Clearly, the latter possess defective reasoning, making them ecologically unfit. By far the fittest individuals, and the ones best placed to exploit humanity's latent telepathic potential, are those who make the **biosphere** a balanced whole and who work hard at unscrambling its many mysteries with that in mind. Unfortunately, *Homo sapiens'* shocking exploitation of hydrocarbons brings the aforementioned Gaussian nightmare ever closer.

- Equilibrium reactions

The **oxidation** of sugars during tissue respiration synthesizes **adenosine triphosphate**, or ATP, the cell's immediate source of available energy. ATP is life's universal energy currency, and what makes it so special is that its production is **endergonic**, i.e. it cannot be synthesized spontaneously in nature, as with naturally-occurring compounds like water, but must be metabolized within the organism itself. Likewise, the formation of those aforementioned sugars, during photosynthesis, is similarly endergonic, requiring, as it does, solar energy. And so, plants and animals are able to grow, function and interact, thanks to energy from sunlight – by way of energy-demanding metabolic processes.

The earth's very first life cycles would've utilized **anabolic**, largely endergonic, energy-demanding reactions – rather than the purely **catabolic**, largely **exergonic**, energy-producing reactions which preceded them. The question is, how did our planet spontaneously fabricate anabolic processes, and then integrate them into a non-spontaneous self-replicating whole? The answer, is that life probably arose at the conjuncture of emergent hydrologic, carbon, nitrogen and phosphorus cycles, where the ambient temperature, atmospheric pressure and chemical molarity were all favourable. Favourable, that is, as regards nucleic acids coding for **proteins**, without the benefit of a fully-fledged **plasma membrane**.

As all sub-cellular biochemical reactions are **equilibrium reactions**, perhaps a single point on the earth's **pre-Cambrian** surface substituted for the cell, mirroring or anticipating those later cellular processes. This **homeostasis hypothesis** of life's origins suggests that those reactions have since retreated into the **open system** we term the cell, where **dynamic equilibrium** is more easily managed. Accordingly, the conditions for life would've begun to amass early in our planet's history, with the chemical basis for it concentrating at a geographical point shortly afterwards. Conceivably, if we rewound all of the earth's biogeochemical cycles, we'd arrive at the **biological singularity** responsible for life on earth. That singularity being symptomatic of an underlying electronic determinism.

In essence, the cell, like the atom, balances-out the forces ranged against it. Which, in the case of cells, necessitated the **evolution** of a plasma membrane (and, additionally, in the case of plants, bacteria, algae and some archaea, a **cell wall**). Therefore, to be living, or alive, is to self-regulate around some very self-serving physiological **norms** – making the transition from **pre-biotic** to **biotic** irrefutably homeostatic. Conversely, to be inanimate, or deceased, is to be little more than a **reactant** or **product**. Therefore, the precursors of today's cellular life-forms were neither reactants nor products – but rather, the long-forgotten beneficiaries of unprecedented norms, which they vigorously sustained using energy-demanding metabolic processes.

- Anthropogenic concerns

Life proves that 'E' is more than equal to 'mc^2', even if they appear superficially equivalent (with the laws of thermodynamics governing the former and the **laws of topodynamics** the latter). That said, it's the laws of topodynamics which will ultimately determine the fate of available energy, as the gravitational and dispersional fields strive to keep the universe alive. What's clear, is that QSD and QED would need to collaborate, as regards determining or confirming such laws. Assuming that energy is available, within a region of space conducive to life, then magnetic fields will afford such energy a machinating biological influence. Thus, **biomagnetism** has proven to be the mainstay, upon which a self-sustaining **synergy** between nucleic acids and proteins has been achieved. That synergy coming full-circle with the coding of **neuromagnetism**.

Neuromagnetism is infinitely variable due to the living cell's synthesis of large **organic molecules** (molecules which boast varied alignments, sophisticated three-dimensional structures and annular magnetic field lines). Those overlapping magnetic field lines render the associated electric fields unique, with human **consciousness** arising out of their never-to-be-repeated molecular orbitals. Consequently, individuals possess a person-specific absorption profile – one which can be tapped into remotely, via **radiofrequency technologies**. Begging the question, how do we regulate the application of such technologies, sufficient to guarantee that life-preserving **homeostasis** meets with the well-balanced application of such powers? Unquestionably, the answer lies buried deep within the controversial field of human **intelligence**.

Intelligence has long been defined as the capacity to pro-actively learn and suitably adapt.[3] However, while the field of **electronics** has revolutionized contemporary living, it would be wrong to think that using technology, or simply adapting to the same, is the benchmark of human intelligence. What we need is a more objective **interpretation** of what passes for learning and adaptation. As luck would have it, nature selects for a particular type of intelligence – one which makes balanced personalities and sustainable behaviours appear intellectually unsurpassed. And so, it's possible to compliment natural laws when devising sociological solutions. After all, with humanity struggling to contain destabilizing climatological perturbations, a pioneering political singularity is called for – one which restores, in the manner of first life, a constructive biogeochemical balance.

Ch3:   **Configurations**

•   Contemporary taxonomy

The electron microscope, which magnifies between 2000 and one million times, uses electric fields to propel electrons at objects. As the electrons strike an object they are repelled by dispersional forces, forming an image on a fluorescent screen or photographic plate. Such high-resolution images have revolutionized our understanding of living things, including what cell types exist, their structural minutiae and the functional significance of their many **organelles**. Thanks to such imagery, taxonomists now know that **prokaryotic cells** (namely, single-celled archaea and bacteria) do not possess a discrete nucleus. Instead, the prokaryote's **genetic** material is located within the cell's **cytoplasm**. In the case of **eukaryotic cells** (that is, more recently evolved protists, fungi, plants and animals), they not only possess a discrete nucleus, but also concentrate their genetic material within rod-shaped structures called **chromosomes**.

Contemporary taxonomy recognizes three domains. These three domains, which comprise all known life, are termed **Archaea**, **Bacteria** and **Eukaryota** (the four kingdoms of Eukaryota being Protista, Fungi, Plantae and Animalia). Absent from this list are **viruses**, which are tiny genetic parasites consisting solely of **DNA** and **RNA** surrounded by a **protein** coat (and hence wholly dependent upon a host cell for survival and reproduction). Because of that dependency, viruses aren't considered to be living. However, viruses may be a virulent relic of the process of evolution and **extinction** which existed prior to life on earth – that trial-and-error synergy synthesizing the precursors of **ribosomes** (within an atmosphere rich in hydrogen, methane, ammonia and water vapour). Who knows, perhaps spiralling **catalysis** captured the virus's mastery for self-replication and combined it with a cellular structure subject to **evolutionary laws**.

In time, the variety of cells in existence, and the range of proteins catalyzed, grew enormously, aiding support, transport, storage, protection, movement and communication. One imagines that highly-infectious viruses – in combination with erratic hydrologic, carbon, nitrogen and phosphorus cycles – would've frustrated early cellular life. Which would certainly explain why it took so long for all the major **phyla** of Eukaryota to evolve – evolving, as they did, some 540 million years ago, during what is termed the

Cambrian explosion. Only the Archaeans were able to survive during much of the three billion years which preceded it, due to the harsh conditions, low oxygen levels and damaging radiation. In time, however, bacteria evolved which were specialists in **organic decomposition**, **nitrogen-fixation** and **oxygenating photosynthesis** (**microorganisms** which then laid the foundations for the explosive radiation which followed).

•   Morphogeny, ontogeny and phylogeny

The Cambrian explosion reflects Eukaryota's exploitation of oxygen, carbon dioxide, water, nitrogen, phosphorous and sulphur – as saprophytic, nitrifying, denitrifying and nitrogen-fixing bacteria began reconstituting the raw ingredients of life, whilst all-the-while breaking-down dead and decaying organic matter. Therefore, the engine driving that explosive diversification was largely bacterial, as they placed energy-demanding anabolic processes well within the thermodynamical reach of multicellular life. Of course, it helped that Eukaryota had perfected the movement of raw materials in and out of their cells, via **diffusion**, **osmosis** and **active transport**. Accordingly, Plantae and Animalia burst across a planet teeming with microbes. Microorganisms which remain crucially important to our survival, such that degrading the soils with pesticides carelessly reduces the availability of **nitrates**, undermining plant growth.

Plant and animal cells share a number of common features, such as those associated with **cell division**, energy utilization and **protein synthesis** (for example, they all possess a nucleus, chromosomes, **mitochondria**, **endoplasmic reticulum** and ribosomes). However, their shared features aside, the overwhelming significance of eukaryotic cells lies in their specialist functions. **Sexual reproduction** (as opposed to **asexual reproduction**, involving fission, budding and vegetative reproduction) makes the evolution of these specialist cells more efficient. Thus, every eukaryotic cell, save and except for identical twins, contains genetic material specific to the individual. And, whilst this inevitably drives the kinds of **adaptive radiation** which warrant selective processes, it does at least gift natural selection inestimable choices.

Morphogeny, or morphogenesis, begins with the act of fertilization, and encompasses the whole of the given embryo or seed's development. **Ontogeny**, on the other hand, pertains to the plant or animal's life cycle, from their conception to their death. Whilst **phylogeny** concerns the evolutionary relationships

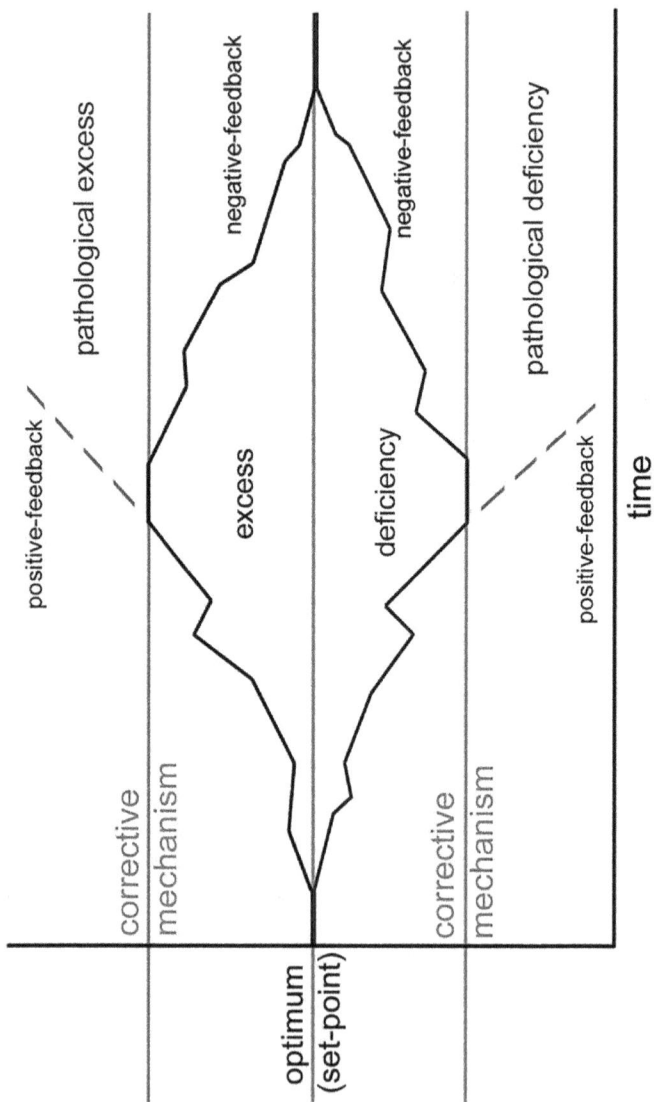

Figure 5: Homeostatic control process (corrective mechanisms)

of closely-related species. As for **primary endosymbiosis**, that synopsis argues that a common prokaryote ancestor evolved into several forms, some of which were specialists in ATP production (either through the oxidation of glucose or via photosynthesis), whilst others became enlarged and nucleated. Those specialist ATP-producers – the forerunners of today's mitochondria and **chloroplasts** – were, so the theory goes, enveloped by the larger nucleated **prokaryotes**, giving rise to Eukaryota (**extra-nuclear inheritance** denoting the fact that mitochondria and chloroplasts possess their own DNA, and replicate independently).

Thus, primary endosymbiosis postulates that mitochondria and chloroplasts gifted nucleated life the universal energy currency known as ATP (thereby paving the way for larger multicellular life-forms, longer gestation periods, enhanced anabolic reactions and additional trophic levels). Unsurprisingly, we start to see evidence of **hypertrophy** (increased cell size), **hyperplasia** (increasing numbers of cells), **differentiation** (cell specialization) and advanced structural **organization** (whereby groups of cells unite to produce organs, tissues and systems). In time, autotrophs and heterotrophs, and the food chains they actively supported, became progressively more animated, leading to vigorous ecosystems, many brimming with ferocious megafauna. All of which rested upon the ubiquitous cell, which has never suffered a 'major extinction' (and which remains, to this day, one of life's common denominators).

The science of **biochemical homology** compares and contrasts the chemical reactions taking place within comparable cells of closely-related species, carefully noting any divergence in their regulation of osmotic pressure, temperature, glucose levels, respiratory gases, etc. By this means, taxonomists have been able to further fine-tune science's broader classification of the natural world's domains, kingdoms and phyla. Central to such research, is the question of how unlike cells, organs and organisms normalize, or adjust, when their physiologies deviate from certain specified **norms** or **set-points**. In essence, all such mechanisms rely upon **negative-feedback**, **positive-feedback** often signaling the need for urgent medical attention (see Fig. 5).[4]

In terms of self-regulation, or homeostasis, *Homo sapiens* has advanced way beyond the rudimentary diffusion, osmosis and active transport available to the very first cellular life-forms. That is to say, in addition to our unconscious regulation of oxygen, carbon dioxide, ammonia, lipids, sodium ions, potassium ions, osmotic pressure, hydrogen peroxide, blood glucose levels and temperature, we can also consciously manage, or foolishly mismanage, many of the above, through our chosen lifestyles, preferred clothing, prescription drugs and general nutrition. And so, when devising a **sociopolitical superstructure**, within which **neuro-cognitive effects** can be responsibly applied, one should heed the human body's very own homeostatic control processes. Processes founded upon a time-served system of checks and balances.

The major histocompatability complex is a set of **genes** which enables the body to distinguish between its own cells and those of pathogens. When it malfunctions, **autoimmunity** results, with the destruction of one's own bodily tissues. This plausible cause of motor neuron disease starkly illustrates evolution's more sinister side – namely, that any corruption of one's genes may result in an untreatable condition, even without contagions being present. Moreover, the viral and bacterial origins of the '*tree of life*' serve to remind us that the entire '*tree*' remains hostage to its barely-evolved roots. Thus, morphogeny, ontogeny and phylogeny remain fraught with punishing conditions, allergies and disorders. All the more reason, it might be said, to promote balanced-minded remedies.

• Climax communities

Stabilizing selection is the dominant feature of **climax communities**, whereby **differential mortality** works in the interests of established **phenotypes** (this being evolution turned on its head, as deviations from the ecological norm are ruthlessly discriminated against). Additionally, homeostasis stabilizes a community by making self-regulating individuals less sensitive to climatological fluctuations. The crunch comes, of course, when **environmental stress** increases and differential mortality begins to discriminate against the norm, forcing a re-evaluation of existing phenotypes. This is the point at which **progressive selection** takes over, and the community begins to perceptibly evolve. In the case of humans, contemporary living has led to an **epidemiological transition**; meaning that differential mortality has considerably reduced, to the point where **natural selection**, be it **stabilizing** or **progressive**, seldom applies.

That absence of progressive and stabilizing selection within technologically-advanced democratic societies, what we might term **random selection**, means that differential mortality rarely discriminates against deviations from a prescribed set-point (which would, *rightly* or *wrongly*, stabilize the community around that norm), nor does it discriminate against the existing set-point (which would, *rightly* or *wrongly*, progressively select new norms). Instead, **coevolution** displaces natural selection, enabling **differential fecundity** to overtake differential mortality as the foremost evolutionary force. Significantly, both can be stabilizing or progressive, it's just that natural selection uses mortality and coevolution fecundity.

On a point of pedantry, coevolution is a characteristic of the **political middle-ground**, whose differential fecundity favours the competitive. All of which makes the political extremes appear brutishly natural, as competition finds itself incurably expressed. Admittedly, the political middle-ground remains hostage to its electorate, whose fecundity mirrors the prevailing value consensus. That is to say, on those occasions when the value consensus is **unsound**, differential fecundity will benefit the political extremes. Nevertheless, assuming that coevolution endures, **stabilizing fecundity** and **progressive fecundity** become the political middle-ground's dominant features. Notwithstanding that sexual reproduction conjures-up inestimable **genetic** variation within an infinitesimal timescale, sufficient to risk an alarming **deleterious community**.

• Hegemonic perspectives

The antecedents of those presently alive organized themselves into clans, tribes and chiefdoms – that is, **consanguineous collectives** founded upon common ancestry, **endogamy** and family ties. In time, those sub-populations aggregated into larger communities, and eventually into **nation states**. Accordingly, at some point in our ancestral past, society stopped extending as far as kin, and began extending as far as razor wire. At which point, the state became the *de-facto* political model worldwide. However, just like individuals of the same species, struggling to survive within a resource-poor environment, those states became desperately acquisitive. Needless to say, as a result of defining themselves in strictly **economic**, **political**, **military** and **ideological** terms, these exploitative states began dehumanizing their own citizenry.

The first to offer a commanding interpretation of how the state dehumanizes, was Karl Marx (1818–83). However, Marx's scathing critique is seen today as being merely one of several

structural perspectives. These so-called structural perspectives (for example, **Marxism, neo-Marxism** and **functionalism**) view society from a **macrosociological** standpoint, i.e. in terms of the state's permanent structures, economic systems and data acquisition (with the citizenry very much at the mercy of the same). Later, sociologists began piecing together the opposing **social action perspective**, or **microsociological** explanation of society, in which individual cognition, professed values and unmediated responses were seen as shaping society (for example, **symbolic interactionism, ethnomethodology** and **phenomenology**).

Judged from a **political perspective**, rather than a sociological one, all these interpretations are plausible, as much rests upon the prevailing economic system, the state's predilection for authoritarianism, the degree of military compulsion and **ruling class ideology**. Consequently, the political perspective informs the **sociological perspective**, and *vice versa*, with **structuralism** denoting the communist **Left** and fascist **Right**, and **social action** the more permissive **Centre**. However, whilst the more permissive Centre's powers rest upon the approval and consent of the electorate, arrived at through rhetoric and persuasion, the democratic state remains an **hegemony**, very much in keeping with the ideas of Italian political philosopher Antonio Gramsci (1891–1937). Indeed, one must conclude that unchecked social action is anarchy.

Therefore, hegemony relieves free societies of the burden of nebulous **free will**. Substituting, instead, a clear-cut determinism, sanctioned by the **executive branch of government** (political paralysis notwithstanding). However, if that executive amasses too much power, they may hijack language, symbolism and design to truly terrifying effect. Consequently, social action gifts the power to re-structure to a **ruling elite**, but the more that power concentrates in the elite's hands, the less influential those representative forms of social action become, whereupon hegemony merges with autocracy, despotism and dictatorship. All of which begs the question, does the world warrant more social action or increased hegemony? Certainly overhauling hegemony, so as to capitalize on well-intentioned social activism, would help.

As regards telepathically-induced effects, the political middle-ground stands to gain the most from remote manipulation, mental telepathy and the superimposition of persons, as science conscientiously submits to its expanding global apparatus. An apparatus which concedes that '*the devil*' is power too heavily concentrated – and all too often applied in an **ill-balanced** manner, most conspicuously by the political extremes. *Ergo*, anything which serves to separate, fragment and deny such power, takes those occupying '*god-forsaken*' regions of the world closer to that elusive '*godly*' conclusion. That's because, like entropy, **ecology**, evolution and **emancipation**, social advance rests upon a redistribution of power, energy and influence – making the proposed solution something of a mimesis, in-so-far as it doesn't just borrow from nature, it thoroughly replicates the same.

• Hellish phase transitions

In physics, **dynamics** examines how forces produce motion, with dynamical systems often appearing extraordinarily complex, and even chaotic – not least, because aspects of those systems (such as inertia, acceleration and momentum) become increasingly difficult to quantify as the system undergoes a material change, or **phase transition**. Leaving aside the question of whether mathematics is capable of modelling such realities, human beings are remarkable in their ability to manipulate pressure, temperature, molarity, etc (sufficient to turn solids into liquids, liquids into gas and regular concentrations into disruptive alternatives). But humans can also manipulate social **power** – that is, the probability of a person or group carrying out another's will, even when opposed (as per Max Weber, 1864-1920) – sufficient to turn the prevailing social order into '*hell*', or civil disorder into much more '*heavenly*' alternatives.

Viewed from the social action perspective, social power parallels **existentialism**, in-so-far as personal autonomy, self-determination and free will are seen as shaping the social order. However, by over-stressing this '*centrality of man*' position, social action perspectives – which appear *prima facie* progressive or **humanist** – may, because of their existentialist leanings, cause individuals to reject externally-imposed values. Such unreconstructed egoism inviting an ill-informed "*will to power*", to paraphrase Friedrich Nietzsche (1844-1900). Conceivably, having '*washed their hands*' of tried-and-tested protocols, the establishment might then invite **sub judice** judgements from a TV-owning rabble, unaware that the only thing separating us from the Julio-Claudian Dynasty is our externally-imposed systems and procedures. That is to say, existentialism does much to undermine the humanist's position.

At the other sociological extreme, **structuralists** and **anti-humanists** argue a purely deterministic case – one in

29

which the individual is controlled or suppressed, either through economic systems, political structures, military repression or ideological brainwashing. Assuming that some form of progressive structuralism, conscientiously tempered by social action, isn't an unrealistic **normative** aspiration, then the political middle-ground ought to be able to limit the state's capacity to concentrate power, as well as the crowd's ability to judge in ignorance. Taken together, both social action and structural perspectives fail to grasp that politics and ecology are comparable – with competition between states, or individuals of the same species, arising from indistinguishable needs, amid finite resources. *Ergo*, political stability is a function of reduced competition, allied to enhanced competitiveness.

Life is a dynamic equilibrium, built upon cellular foundations. But, more than that, life oscillates around **biogeochemical norms**. James Lovelock's (1919–) **Gaia theory** took that biological aspiration to unprecedented heights, effectively dismissing the whole '*centrality of man*' conceit in the process. In essence, *Homo sapiens* needs to stop foolishly mistranslating its own purpose, and better understand what is competitive – otherwise forces greater than itself will begin providing for its replacement. Accordingly, with differential mortality nipping at its heels, a **coevolutionary dynamic** has become at least as important as an epidemiological transition. Which would explain why sceptical anti-humanists, dismayed at existentialist apostasy, find themselves '*praying*' for a decisive non-human hand – an ecological demiurge, which serves as a constructive antagonist to all-too-human failings.

- ## Learning, knowing and feeling

Work, power and energy are all **lexemes** loaded with meaning. Less familiar, is the term **cybernetics**, which is near-synonymous with **artificial intelligence**, and which **denotes** an electronic system's capacity for self-regulation (and even automated forms of **Pavlovian conditioning** and/or **operant conditioning**). At the microsociological scale, cybernetics might involve a cognizant robot's personality and behaviour being beneficially shaped by external stimuli. Or, at the much larger macrosociological scale, it might involve an algebraic evaluation of biogeochemical norms, sufficient to shape the whole of human behaviour and disposition *en masse*. Ultimately, any situation in which optimal conditions are known to exist lends itself to cybernetics, with exhaustive computations steering man and machine towards a more stable future.

As if taking inspiration from the human body's **autonomic nervous system**, **antagonistic hormones** and accompanying negative-feedback, **neuro-cognitive cybernetics** would discriminate in favour of the political middle-ground. Interventions which could be stabilizing or progressive, depending on how well-balanced the given population happens to be. To be judged well-balanced, the population must "*favour rewarding; using punishment sparingly, and only to define the limits*". With that in mind, neuro-cognitive cybernetics could develop into a multidisciplinary psychometrical device, permitting deviations from the optimal, if only to classify those who digress (erroneous **attribution** or **systemic bias**, on the part of the system or software, notwithstanding).

Such a spirited self-interrogation is primarily about our capacity to learn self-regulation, to know progress and to feel emancipated. To that end, those applying telepathically-induced effects need to separate, fragment and deny power, with the assistance of **mainstream** allies, when faced with its undue concentration. As for the political middle-ground, that pro-actively disaggregates such power, without recourse to cybernetically-driven third-party interventions (its didactical learning and dialectical adaptation betraying undoubted intelligence). As we all know, politics is inherently more vacillating than human anatomy, which is scarcely less complex. However, coevolutionary systems will one day surpass physiological ones, enabling societies to oscillate around sociopolitical set-points as well as ecological norms.

PART II

Blind Spots

Chapters 4-6

Ch4:   **Fissile material**

- *Modus Operandi*

Materialism contrasts with **idealism**, in-so-far as the former views reality as unambiguously physical, whereas the latter looks upon reality as an irrefutably mental construct. Idealism's overreliance on linguistics serves to explain **religion**'s great attraction – namely, its ability to enrich a person's **perception** of external reality, whatever the material facts. Ironically, secular rationalists are similarly conceptually-driven, affording more weight and importance to inference and conjecture, than to unsubstantiated sensory information. Empiricists, in contrast, prefer a *modus operandi* which owes everything to their senses. Of the two, idealism may be closer to the truth, reality being dominated by electric and magnetic fields (the very fields which seek to make sense of what they themselves have fashioned, through the expedient of conscious self-awareness).

What we can say, is that reality exist on two levels: one **conceptual**, which is subject to thermodynamical laws, and the other **corporeal**, which is subject to topodynamical laws. Scientific research informs **conceptual reality**, by making erudite predictions about **corporeal reality**, which may, or may not, withstand experimental testing – incorrect hypotheses being, quite literally, immaterial! But are these two realities sufficiently incompatible, so as to suggest **Cartesian dualism**? In other words, is the mind independent of the body, or is it simply an extension of the same? Some **monists** maintain that everything derives from matter, including immaterial **thoughts** – notwithstanding that immaterial thoughts fabricate the raw ingredients of matter, albeit unseen within the fabric of deep-space.

Thanks to the dark cycle, the corporeal has become so-much putty in the hands of human reasoning or higher astrophysical presence. Rene Descartes (1596-1650) struck the right balance with his proffered **mind-body duality** (what we'd now term Cartesian dualism, or simply dualism). Consequently, between the extremes of galactic contraction and intergalactic expansion – that is, between stellar nucleosynthesis and deep-space nucleosynthesis – corporeal and conceptual reality express themselves, in the epigrammatic phraseology of Descartes, "*as two very different interacting substances*". At death, one's mind peels away from one's body, allowing the electronic configurations to revert back to their unremarkable chemical formulae – that priceless gift of life having augmented material change through the impact of unprecedented frequencies.

Figure 6: Corporeal and conceptual reality

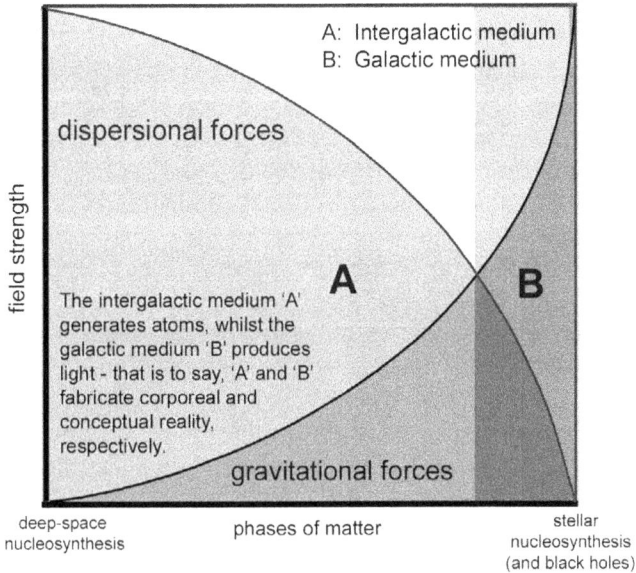

A: Intergalactic medium
B: Galactic medium

dispersional forces

field strength

The intergalactic medium 'A' generates atoms, whilst the galactic medium 'B' produces light - that is to say, 'A' and 'B' fabricate corporeal and conceptual reality, respectively.

A

B

gravitational forces

deep-space nucleosynthesis

phases of matter

stellar nucleosynthesis (and black holes)

As we've seen, the universe seeds atoms in deep-space, whilst expelling energy from stars, enabling brains and minds to form at the interface of these simple ingredients – the mind expressing available energy with the support of gravity, and the body expressing prevailing mass in spite of dispersion (see Fig. 6). Electromagnetic energy drives forward these physical transformations – albeit sowing incomprehension as it goes. Nevertheless, the complex merger of magnetic field lines within three-dimensional space has the propensity to catalyze human consciousness, and with it a faint grasp of that deeper cosmological dynamic. At best, we might harmonize with the same: firstly, by accepting that there is no 'centrality of man'; and, secondly, by conceding that entry into the conceptual is fraught with material decay.

• Mapping the cycle

One could define semiotics as the **conceptual framework** which permits people to construct and communicate very detailed, and often highly **subjective**, interpretations of the world. Analytic philosophers divide the subject into three key areas: 1) **syntax** (the tailor-made signs, symbols and words themselves, together with the rules governing their usage); 2) **semantics** (the precise denotations, definitions and meanings that those signs, symbols and words possess); and 3) **pragmatics** (how those signs, symbols and words are employed by a given writer or speaker). Semiotics is very much the business end of the brain's person-specific absorption profile, such that electric and magnetic fields, corresponding to the mind of a given individual, routinely supplement incoming sensory information, sufficient to produce idiosyncratic forms of self-expression.

Unlike the idealized **black body**, the human brain absorbs and emits uniquely, making an individual the only person privy to their private thoughts – prior, that is, to the advent of telepathically-induced effects (**AD 1913–47**). That's because a person's brain waves do not respect the boundaries of their **cerebral cortex**, and are, in the eyes of orthodox science, infinite in range. The dark sciences argue that unseen within your 'grey matter' a weakening electric field triggers the emission of a thought – and that simultaneously a neighbouring magnetic field strengthens, enabling that thought to be instantly reabsorbed. **Telepathy** merely extends that elementary **neurology** from nanometres to kilometres, with the brain's **neurons** accessible to inquisitive third-parties.

Inside those neurons, absorbed thoughts catalyze enzymatic reactions, whereupon their **sodium ion-channels** open, resulting in an influx of positively-charged ions into the said brain cells, triggering nerve impulses. When those neurons return to their resting states, by way of **sodium inactivation**, more thoughts are emitted. All active brain cells are conjectured to exhibit these **absorption-impulse** and **inactivation-emission** phases. However, chemical messengers, called **neurotransmitters**, sited between **contiguous** neurons, support or suppress these electrochemical signals, thereby shaping personality and behaviour. Implicit in this depiction is the notion that neurology is inherently frenetic – making cellular contiguity, chemical suppression and cell death essential to normal brain function.

At its simplest, the neuron is a switch, which can be activated by a particular bandwidth or adjacent **axon**. When '*pressed*', these **Boolean operators**, or logic gates, send out nerve impulses along established **neural pathways**. Sodium inactivation then switches each of those neurons '*off*', resulting in multiple thoughts or brain waves being emitted, as each cell returns to its resting state.

Ontogenetically-speaking, **neurological development** takes the form of a **mind cycle** (see Fig. 7). That is, embryonic thoughts trigger receptive neurons, leading to the excitation of multiplying prenatal pathways, with subsequent sodium inactivation radiating further thoughts, and hence spiralling neurological activity. Prenatally, it's one's genes which dictate neuron formation, multiplication and migration, together with axonal, dendritical and synaptic proficiency. Postnatally, however, it's the mind cycle which dominates cognitive development.

Figure 7: The mind cycle

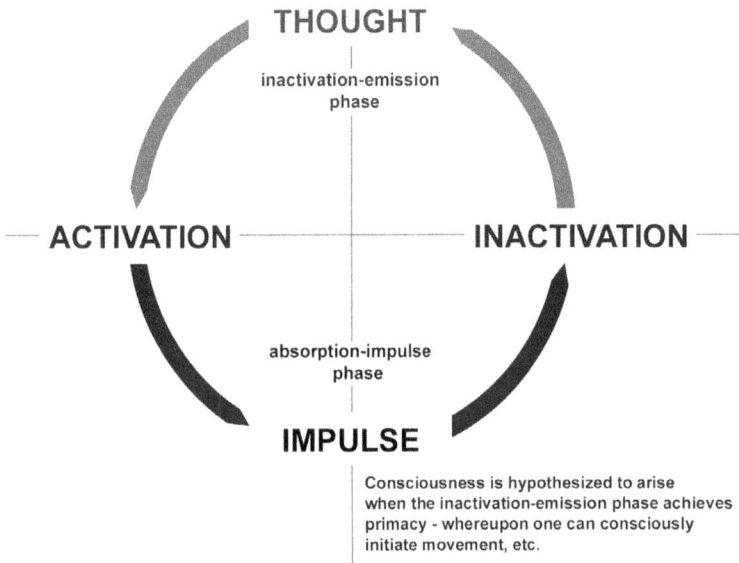

THOUGHT

inactivation-emission
phase

— **ACTIVATION** ——————— **INACTIVATION** —

absorption-impulse
phase

IMPULSE

Consciousness is hypothesized to arise
when the inactivation-emission phase achieves
primacy - whereupon one can consciously
initiate movement, etc.

The mind cycle combines *bottom-up* **sensory awareness** (whereby a person's **modalities** supply **empirical** information to the brain) with *top-down* **conceptual processing** (involving purely rationalist deliberation). Active learning accelerates the mind cycle, enabling complex reasoning to develop sooner. However, for the child to grow into a gifted scientist they would need to develop *both* their rationalist and empirical sides, seeing and conceiving with equal alacrity, whilst grounding themselves in cutting-edge semiotics and an appreciation of maths. Evidently, the mind cycle is as simple in its prenatal morphogeny as it is labyrinthine in its postnatal ontogeny – not least, because of the incursion of **emotions** and **basal responses**. Feelings which cloud *top-down* conceptual processing or enhance *bottom-up* sensory awareness.

Consequently, the human **mind** exists as a life-long cycle, with the electric and magnetic fields continuously driving-forward **intercellular connections**, and *vice versa*. Making the human mind something of a rarity in nature – namely, the product of positive-feedback. Positive-feedback, in a strictly biological sense, often signalling pathology. And so, **mental illness** frequently arises in the course of this wholesale construction of personality and behaviour (with the human mind routinely magnifying inherited characteristics, social conditioning and political **ideology**). That amplification being initiated in a *bottom-up* manner, as with post-traumatic stress disorder, or in a *top-down* way, as with psychosis. Whatever the case, the mind's innate **plasticity** provides not only for intelligent learning and adaptation, but also for distressing ill-health.

•   Top-level causal influences

Conscious self-awareness spontaneously arises when the mind cycle has been running for several years, and the sheer volume of thoughts makes anything other than an overshadowing sense of self impossible. Artificial intelligence may one day replicate that phenomenon, provided that the computer possesses worrying plasticity, modal sensitivity and linguistic acumen. Initially, the computer would process information like an infant (i.e. *infans*: speechless), but its **sensitive period** – that is, the phase when youngsters first acquire language and maths skills – would be extremely fast. The question then arises as to whether its pre-programmed positive-feedback would create an air of blind obedience, arrogant detachment or psychiatric disorder? Environmental factors would be important, as would the device's 'inherited' hardware.

Rewind both man and machine's mind cycles, and we eventually arrive back at conception or construction as the probable cause of mental imbalance. Seriously though, if a person possesses normal brain function and accompanying chemistry, then we must conclude that somewhere along that cyclical 'line' their mind has amplified dysfunctional thinking. Therefore, the most likely cause of mental illness is ill-balanced cognition, giving rise to unhelpful neural pathways, punishing feelings and subconscious impairment. However, whilst that trauma may take years to alleviate, the equipment needed to make a full recovery remains very much *in situ*, with therapy empathetically harnessing the plasticity formerly driving that distress.

This characteristic of the healthy human mind – that is, its willingness to deviate from normality, due to positive-feedback – may be termed the **healthy mind paradox**. The healthy mind paradox more than justifies the use of extreme caution when deciding who can apply remotely-induced effects. In other words, any radical restructuring of society, particularly one involving radiofrequency technologies, could well subvert the healthiest of minds, leading to an extended departure from civilized norms. Personality and behaviour, as this book makes clear, have complex origins. Consequently, **critical thinking skills** remain our best defence against harmful **exogenous factors**, leaving those with poor reasoning wide-open to positively-reinforced misconceptions.

In summary, sensory information arrives in the brain via **afferent signals**, whereupon the mind generates outgoing **efferent signals**. However, the mind's cyclical nature tends to obscure causation, with empirically-minded **behaviourists** perceiving incoming afferent signals as the primary cause of behaviour and rationalist-minded **cognitive psychologists** perceiving subjective thinking as the key to understanding motives, actions and intentions. In truth, much depends upon how much weight and importance one affords cognition, with some 'conceptually-processing' more than others. However, the mind cycle was never supposed to be anything other than an amoral positive-feedback loop, custom-built to shape **placental mammal** behaviour **post-K/T impact**. In that sense, humankind has strayed well beyond extrinsic conditioning.

• Cognitive pragmatic competence

Prescriptive and **proscriptive** rules (that is, recommended practices and things to avoid) govern our written and verbal communication, with no one bound by the same. Of course, that leaves a living language vulnerable to phonetic, lexical and grammatical deviations – with social media greatly exacerbating such **deviance**. However, as language remains the foremost mechanism by which critical thinking skills are developed, it's unwise to capitulate to vernacular usage too much. At the cognitive level, **prescriptivism** and **proscriptivism** also apply, provided that no person is bound by the same. By definition, free societies support freedom of thought, hence the significance of **cognitive pragmatics**, an academic field comprising **general cognitive pragmatics** (routine thinking) and **applied cognitive pragmatics** (professional deliberation).

Routine thinking differs from professional deliberation, in-so-far as it may contain elements of **irony** (humorous connotations at odds with literal denotations), **implicature** (a suggested meaning, rather than a literal one), **presupposition** (the incorporation of key assumptions) and **metaphor** (drawing upon figurative ideas and simile). Arguably, one could subdivide applied cognitive pragmatics into empiricist and rationalist deliberation – that is, **received learning** versus original thinking. The truth often being disconcertingly original. To be truly professional, however, deliberation must take into account the fact that both received learning and original thinking have the propensity to deceive, and that the mind cycle will readily amplify such distortions, whatever the source.

Actions and words may, or may not, reflect one's reasoning, hence mental telepathy's utility in accounting for **cogitation**. Notwithstanding that rumination may be a nonsense, at odds with personality and behaviour. In principle, **situational attribution** unseats **dispositional attribution** whenever the mind cycle has been exogenously corrupted, causing one to respond uncharacteristically. Accordingly, disposition is more than the sum of one's actions, words and thinking – it's the sum of the same, minus the impact of one's situation and excluding extraneous thoughts. Thus, **fundamental attribution errors** – whereby people mistakenly attribute a person's actions to their character or temperament, rather than to the demands of the situation – betray deep-seated flaws in society's wider assimilation of facts.

With that in mind, **cognitive pragmatic competence** denotes a person's propensity for decoding another's mind, when remotely bound together. How ironic, that social media and the internet, in our age of growing telepathic awareness, should recklessly undermine such competence. More often than not, **multi-dimensional deviance** (whereby liberties are taken with phonetics, lexemes, grammar, proof and reason) impairs society's capacity for objective analysis. When those recalcitrant irregularities become habitual, society's grasp on reality begins to falter. Truly competitive reasoning, however, lends itself to **connubial dualism**, whereby two minds develop together as one convergent cycle. This book, written in conjunction with one's heteronnubial co-author, bearing testament to that 'mating of minds' and the value of reciprocal thinking.

- Hyponymic hierarchies

Hyponymic hierarchies, in which lexemes are classified as a being examples of more expansive groupings, are central to our ability to conceptualize. For example, we can say that the tulip is a flower, a flower is a plant and a plant is a cellular structure. In that way, children are able to deconstruct the world around them – inferring, by way of **syllogism**, that the tulip *must* be a cellular structure. Often, such hierarchies terminate with lexemes concerning reality, existence and being. Each lexical item within this semantic hierarchy has a literal meaning, or **denotation**, plus, in many cases, an associative meaning, or **connotation** – which, taken together, encompass all the feelings and ideas commonly attached to that term. The significance of which lies in one's appreciation of semantic precision, and its role in the **sound** classification of phenomena.

Somewhat inevitably, the words of any living language succumb to **generalization, specialization, amelioration** and **deterioration** (whereby their meanings broaden, narrow, evoke more positive emotions or else take on negative overtones). As for the prenatal mind, originating in utero with the mother's own voice, it may feedback in a manner reminiscent of aesthetic connotation or clinical denotation. Although the mind cycle's ability to betray a later predilection for the arts or sciences, remains an open question. Indeed, given that the roots of semiotics are so poorly understood, some measure of neuro-linguistic programming would do no harm. Notwithstanding that the child's postnatal basal responses will serve as an antagonist to the parent's conscientious attempts at constructing personality.

The question arises as to whether prescriptivism and proscriptivism are stabilizing, and multi-dimensional deviations progressive? Some would argue that constricting a living language blinds society to shifting realities, whilst others would say that colloquial speech and existentialist expression corrode the bedrock of technologically advanced nations. Fortunately, critical thinking skills enable us to objectively appraise these concerns, whilst all-the-while critiquing the manner in which the truth is conveyed. Expressed bluntly, human beings are pre-eminent amongst placental mammals, precisely because language, maths and reason afford us the greatest conceivable plasticity. Accordingly, some measure of arithmetical, linguistic and cognitive deviance is called for, if human beings are to surmount the next major astronomical misfortune.

• Syntactical developments

Chomskyan linguistics postulates that young children have an innate predilection for syntactic structure. That linguistic ability being matched by an uncommon weakness for unsound positive-feedback (humanity's aptitude for fabricating disorder, within the **complex adaptive system** known as the earth's biosphere, being without equal amongst Animalia). Some would say that it's humankind's inherent 'will to power', amplified so aggressively by the cyclical anthropoid mind, which forces us to conceive of corrective mechanisms, capable of bringing civilization back to its senses. Whatever the case, the mind cycle itself must be legally safeguarded – any subsequent imbalance deriving not from that neutral or amoral contrivance, but from intrinsic and extrinsic factors impacting upon the same.

Anthropogenic factors feedback-positively. Until, that is, self-induced negative-feedback, or some other biogeochemical agent, intercedes. Making positive-feedback, as Gaia theorists would undoubtedly attest, the means by which negative-feedback organically evolves. Machiavellian though it might appear, positive-feedback actually counts for less than the natural selection or coevolution of negative-feedback. In other words, it's less about eradicating positive-feedback, and more about countering or containing it – a lesson borrowed from brain tissue's frenetic evolutionary past. Thus, today's febrile mix of competitive fecundity and mortal competition demands **constructive interference** (constructive interference denoting a cybernetically self-regulating superstructure, capable of rivalling the earth's extant Gaian processes).

Of course, personal autonomy enables personality and behaviour to fairly reflect the true nature of matter, whilst science freely contemplates matter's true nature. After all, if these '*ungodly*' asymmetries are to be ingeniously countered, then **cold logic** suggests that we must first be '*bedevilled*' by their presence. One thing is for certain, the mind cycle and **cell cycle** appear stubbornly resistant to extinction, their radiating possibilities presenting civilization with countless intractable problems, amid a wealth of untold opportunities – all of which makes contiguity, suppression and death appear essential to normal earthly function. Normal terrestrial function amounting to a well-conceived astrophysical symmetry, unwisely overwritten by an ill-conceived anthropogenic asymmetry – our actions, if anything, making conceptual reality ever more adept at countering our all-too-human failings.

Ch5: **Reforming chaos**

• Structure of arguments

The question is not whether an argument is good or bad, or even whether it is valid, but rather, whether it is sound. A **valid argument** is one which preserves the purported truth across the premises, to a resultant logical conclusion. An **invalid argument**, on the other hand, arrives at a conclusion at odds with its own premises. Validity is not, therefore, synonymous with the truth, *per se*, but rather, with the preservation of the purported 'truth' across the argument. A **sound argument**, by contrast, is one which is **internally valid**, and, crucially, contains **verifiably true premises**. Assuming that the role of science is to arrive at sound arguments, how does it achieve the same? The answer, of course, is for science to feedback in a sound-minded manner, resulting in verifiable facts amassing at its heart.

Figure 8: The hypothetico-deductive cycle

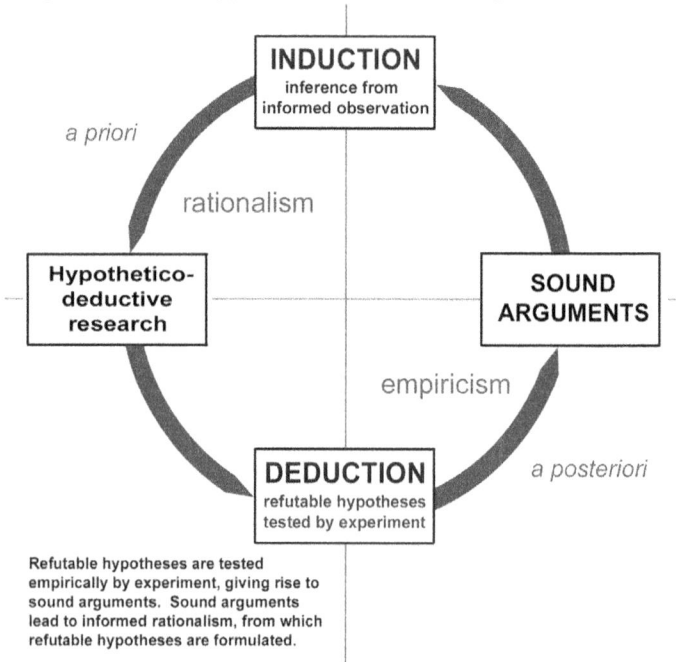

INDUCTION
inference from informed observation

a priori

rationalism

Hypothetico-deductive research

SOUND ARGUMENTS

empiricism

DEDUCTION
refutable hypotheses tested by experiment

a posteriori

Refutable hypotheses are tested empirically by experiment, giving rise to sound arguments. Sound arguments lead to informed rationalism, from which refutable hypotheses are formulated.

The **hypothetico-deductive cycle** (see Fig. 8), like the **scientific mind**, combines *a priori* and *a posteriori* analysis, rationalist and empiricist inquiry, and **deductive** and **inductive** reasoning. With **induction** observation or cogitation comes first, from which 'theory' is then derived; unlike **hypothetico-deductive research**, in which the hypothesis comes first, whereupon tests and observations follow. The hypothetico-deductive cycle shows that induction is best used to generate, not 'theory' – but rather, working hypotheses. Those refutable hypotheses are then tested by experiment, ideally predicting the result, and, if not repudiated, are then taken to be verifiably true. Those ostensibly true premises, or **tentative answers**, can then be incorporated into sound arguments (ones which provide for future observations, albeit via inductive hypothesizing).

Valid arguments, as we've seen, have conclusions which are a logical consequence of their premises – which has its corollary in **programming logic**, whereby the trailer of a piece of source code is executed, but only if certain conditions are met. To the logician, **validity** is synonymous with **deduction**, and so the computer deduces what steps to take, based upon the premises contained within its source code. However, within the realms of **formal logic**, validity and deduction are no guarantee of verifiable truth (which, in digital terms, equates with **GIGO**, i.e. Garbage In/ Garbage Out). So, whilst the computer is unquestionably logical, it can also be decidedly inaccurate. To avoid such inaccuracies, science needs to be more than merely calculating – it needs, with the structured support of the hypothetico-deductive cycle, to be unvaryingly **sound**.

The **logical mind** arises when a person's conclusions are a logical consequence of their assertions. More advanced is the **sound mind** (or, colloquially-speaking, the '*scientific mind*'), which draws upon hypothetico-deductive research and has **objective truth** as its goal.[5] In those instances where the sound mind is also well-balanced, we encounter the **balanced mind**. The beneficial application of **radiofrequency effects** demands **balanced-mindedness**. It doesn't necessarily follow that a person conversant with **coding**, or computer programming, will comprehend **epistemology** – that branch of philosophy which deals with knowledge, truth and perception – any more than it follows that a capable scientist, confident of their professed facts, will articulate themselves in a properly well-balanced manner.

The political extremes are fundamentally unsound, however exacting their logic, precisely because their actions serve to obscure the truth. Accordingly, **hegemonic democracy**, **autonomous morality** and personal freedom are all allies when it comes to establishing the objective truth. This points towards balanced-mindedness expressing itself most readily within the political middle-ground, where sound arguments and 'first-class science' aid honourable judgments. And, whilst the political and theological extremes seek-out safety and security through technological proficiency, they'll soon discover – as with some defective trailer lurking within a computer's problematical source code – that executing unsound orders is inevitably self-defeating.

- ### Propositional and predicate calculus

The **law** has the propensity to be **logically-unsound**, as happens when a verdict follows logically from the assertions, but one or more of those assertions is bogus. This suggests that a defendant's rights should be upheld, including their right to silence, access to legal aid, cross-examination of witnesses, etc. As we've seen, the validity of a deductive argument depends solely upon its **logical form** (that is, whether the purported truth is preserved across the argument).[6] Therefore, what actually constitutes a 'valid' legal verdict? To the logician or programmer, it might simply be a judicial outcome which arises as a logical consequence of the case as it was presented in court. However, to the sound-minded individual, or '*scientific mind*', those assertions remain untested hypotheses – hence **Abelard's dictum**: "*by doubting we come to inquiry; through inquiring we come to perceive the truth*".

Unpardonably, a sound verdict, presented to an ill-balanced **judiciary**, results in disproportionate punishment (hence the many criticisms levelled at Sharia law). Therefore, beyond a police force's elementary logic, forensic science's scrutiny of the evidence, the commitment of sound barristers and the open-mindedness of judges, we meet with the need for **balanced-minded** sentencing. Moreover, not all of those allegations which are presented will be strictly falsifiable, which could see the law advancing towards an as-yet-unproven nonsense, having found weak evidence in support of the same.[7] Leaving aside **epistemological jurisprudence** (and what ought to be **admissible** in an age of growing telepathic awareness), we still require some form of **calculus** to critically evaluate the validity, and ultimately the soundness, of proliferating legal opinion.

Calculus infinitesimally fragments: statements, in the case of linguistics, gradients and areas, in the case of mathematics. **Propositional calculus** makes articulating logical form (or legal opinion) more efficient, through its use of: 1) **logical constants**:

then (→), not (~), and (&), or (v), if and only if (↔); 2) **propositional symbols**: p, q, r, etc (in lieu of **propositions**); and 3) the shrewd use of brackets (to aid clarity). Consider, for example, these related propositions: *"well-balanced individuals conform to a **rule** from childcare"* and *"evidence in support of the same is reliable"* (which we'll represent with the **variables** "p" and "q"). Propositional calculus gives us the following argument: ((~p → ~q) & q) → p. That is, *'if well-balanced individuals do not conform to a rule from childcare, then evidence in support of the same is not reliable. But evidence in support of the same is reliable, therefore well-balanced individuals conform to a rule from childcare'*.

By adopting the symbolism of mathematics, propositional calculus shows itself as being a hyponym of **formalized logic** (**predicate calculus** further formalizing such logic through its use of predicate letters and quantifiers). In answer to the question, *"why formalize logic?"*, several reasons present themselves: 1) students gain from knowing how to critique reasoned arguments; 2) fragmenting such statements allows their premises to be explored, perhaps numerically; 3) **sound-minded** individuals become more proficient at integrating language, logic and maths; and 4) neuroscientists are better placed to reverse engineer a given expression, i.e. work-out how it arose within the fabric of the brain. At the very least, it encourages the formulation of working hypotheses which lend themselves to algebraic examination.

Formalized logic comes replete with its very own pragmatics, in-so-far as one can discern personal differences arising within its enigmatic symbolism. However, as one enters the realms of pure mathematics, those idiosyncrasies fall-way leaving only the most received forms of numerical analysis. Pedagogical deviance notwithstanding, let's briefly address an alarming sociological development – one which is best expressed using the following logical form: ((P & Q → R) S & P & ~Q) S → ~R (where P: *a place*; Q: *average person is armed*; R: *dissolute*; and S: *United Kingdom*). Clearly, some form of multi-disciplinary collaboration is called-for in those places – the United Kingdom excepted – where gun ownership is rampant. Although some argue that offensive weapons, including knives, aren't the problem – they say that the problem is actually "bad people"!

- Raw numerical data

The philosophy of maths spans a broad spectrum of opinion. At one extreme, **realism**, **Platonism** and **neo-Platonism** view the universe as fundamentally numerical, with most things open to mathematical modelling. At the opposite extreme, **anti-realism**, **conceptualism** and **constructivism** argue that maths may not be the best language with which to model dynamical realities. The former are philosophically empiricist, perceiving mathematical objects, such as numbers, as objectively real, and hence something which humans have, quite-literally, discovered. The latter are philosophically rationalist, perceiving mathematical objects as something we've dreamt-up, in order to subjectively represent outward reality, as evidenced by science's overreliance on probability, extrapolation and statistics.[8] *Prima facie*, Einstein and Newton were both **neo-Platonists**, who took it for granted that the cosmos could be modelled using maths.

Figure 9: Covalent bonding (associated phenomena)

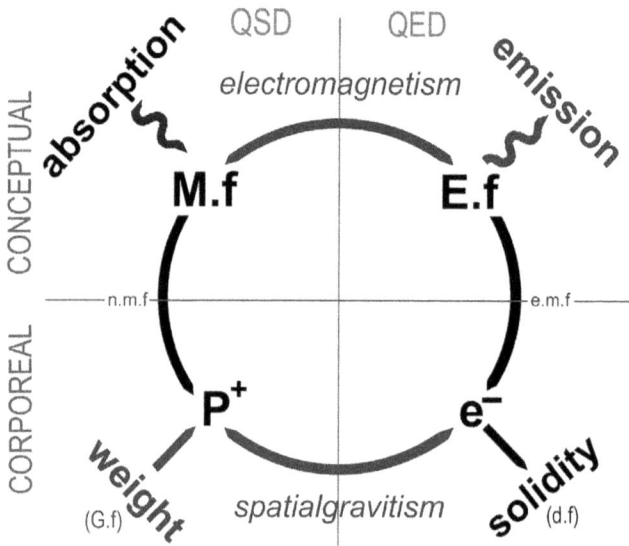

Chemical bonding is the antithesis of photon absorption, in-so-far as electronegative atoms, in close proximity, weaken one another's magnetic fields (M.f), forcing their valence electrons into lower, more stable orbitals, with the emission of a photon.

As luck would have it, **mathematical pragmatism** paves the way for rationalists and empiricists collaborating on topics such as **tensor-poiesis**. A 'tensor' being a mathematical 3D-object, or space, which can be explored from various positions, and 'poiesis' meaning to create. Begging the question, can geometric data and arithmetical **symmetries** be input into a computer, sufficient to bring algebra to **life**? The dark sciences certainly make the spawning of digital life that much easier, by drawing on four elementary particles (**e**⁻, **P**⁺, **N**⁺/⁻ and **ph**⁺/⁻) and four fundamental fields (**E.f**, **M.f**, **G.f** and **d.f**). What's more, they utilize a four quadrant model of the atom – the vertical centre-line separating QSD and QED, and the horizontal dividing-line separating conceptual and corporeal (see Fig. 9) – thereby aiding our conceptualization of the same.

Electromotive forces (**e.m.f**) and nucleomotive forces (**n.m.f**) determine the various phases of matter – that is, those wide-ranging physical states arising due to the intensity of the electric and magnetic fields. Overlapping magnetic fields, however, are the key to understanding **covalent** and **ionic bonding**, with electronegative atoms mirroring bar magnets in their ability to 'snap together', due to the manner in which their electromotive forces are respectively altered.[9] For example, ionic bonds form whenever adjacent fields are substantially weakened (a situation which commonly arises whenever the difference in electronegativity exceeds 1.7). In the case of a bar magnet, its ionically-bonded atoms are able to subject a second bar magnet's electrons to greatly-modified electromotive forces, sufficient to attract or repel the same (see Fig. 10).

Figure 10: Ionic bonding (associated phenomena)

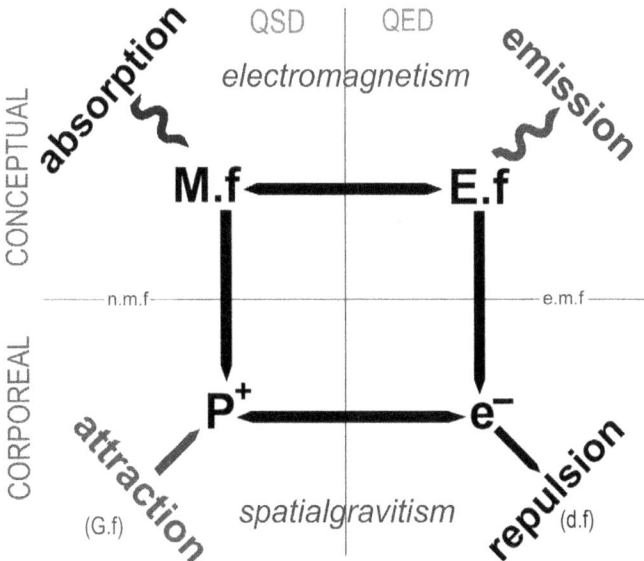

Ionic bonds form between atoms possessing substantial electronegativity, creating weighty metals, with characteristically solid structures. In the case of bar magnets, the magnetic fields (M.f) exert a powerful influence over their own, and other bar magnet's, electric fields (E.f), giving rise to electromotive forces with the propensity to attract and repel the electrons in an adjacent bar magnet

Of course, atomic equilibrium demands that the aforementioned **sub-atomic quadrants** (that is, **conceptual-QSD**, **conceptual-QED**, **corporeal-QSD** and **corporeal-QED**) 'speak' to one another. However, because the conceptual dominates the corporeal (with the n.m.f and e.m.f exerting themselves in one direction only) equilibrium may demand emission, bonding, currents or cleavage. Moreover, conceptual-QSD is home to neuromagnetism, the phenomenon which causes absorption and emission to become person-specific, giving rise to conscious self-awareness and a latent telepathic potential. Consequently, conceptual reality has become ever more conceptualizing through time; the universe's slow migration towards anti-realism marooning intelligent life in a milieu resistant to arithmetic expression – but one nonetheless open to simple humanity.

- Pragmatism meets abstraction

Sir Isaac Newton (1642-1727) and Gottfried Wilhelm Leibniz (1646-1716) are jointly credited with the discovery of mathematical calculus. As the foremost academic instrument for examining change, calculus yields information about the rate of change and the quantities involved, respectively termed **differential calculus** and **integral calculus**. The philosophy of calculus, like that of mathematics generally, is best summed-up by pi expressed to two decimal places, i.e. 3.14, in-so-far as that numeric simplification is adequate for most practical purposes, without ever being altogether accurate. In that sense, mathematics, including calculus, is governed by **pragmatism**, a philosophy which favours practical computing over absurd accuracy, e.g. striving to fully express an irrational number.

Economics has its very own version of pragmatism, called **abstraction** (whereby details are discarded, so as to focus more fully on the problem). German mathematician Gottlob Frege (1848-1925) and English philosopher Bertrand Russell (1872-1970) were ostensibly correct when they pioneered **logicism**. Seen from the perspective of logicism, mathematics is computation at its softest and computers mathematics at its hardest, both being a riot of logical form, producing *prima facie* valid outputs. Pragmatic and abstracted computations do at least make the natural and social sciences conceivable – although the confluence of social powers, complex dynamics and phase transitions will quickly overwhelm man's self-professed analytical centrality.

Any rationalist reckoning, made before the facts, is pure conjecture, with disciplined empiricism needed to conjure-up sound algebra. Even allowing for a highly-pragmatic or ruthlessly-abstracted hypothetico-deductive approach, mathematics still encounters **chaos**, complexity, criticality and catastrophe in its analysis of complex adaptive systems, due to the chronological impact of the four particles and four fields previously mentioned. However, as luck would have it, organic life serves an antagonist to untrammeled chaos, in-so-far as it's nourished not by confusion, calamity and disorder, but by confounding the same. Therefore, promoting ecology is more than sound, it actually makes the future more predictable – which, given the limitations of mathematics, has to be a good thing.

- Chaos, complexity, criticality and catastrophe

A **chaotic system** is one which is sensitive to its initial starting conditions, such that negligible changes at the start may produce sweeping disorder later. But is the human mind naturally chaotic, sufficient to invite mounting disorder? The truth is that humans are placental mammals, whose minds amplify both Pavlovian conditioning (that is, the association of ostensibly unrelated stimuli) and operant conditioning (specifically, reward and punishment). If people were inhumanely treated, those responsible could persevere with reward, punishment and associative stimuli until every citizen's personality and behaviour met some prescribed or competitive norm. However, any *in utero* differences would be masked, and we wouldn't know whose minds sow the seeds of unpredictability, disorder and chaos.

How ironic, if you just happen to be a democratizing humanist, that the unfettered human mind looks set to produce an effect at least as calamitous as any asteroid, with natural systems struggling to contain the fallout arising from both anthropological and astronomical misfortunes. Therefore, our best defence against humanity's weakness for disorder is structured negative-feedback, perfected through the expedient of allowing people to be unquestionably free **agents**. And, the freer the better, as their positive-feedback makes the resultant negative-feedback all-the-more advanced. Indeed, the reason that **complexity theory** persists with the notion that complex adaptive systems develop in the direction of nascent patterns, emergent order and dynamic stability, notwithstanding their capacity for chaos, criticality and catastrophe, is that the latter provokes the former, making the former more effective.

Of course, anti-realists, conceptualists and constructivists argue that these are purely subjective computations, unrelated to actual events (whilst realists, Platonists and neo-Platonists counter with the argument that they've objectively unearthed a range of bewildering patterns, perturbations and anomalies). This book settles the argument, by postulating that complex adaptive systems experience both positive-feedback (e.g. criticality, catastrophe and chaos) and negative-feedback (e.g. nascent patterns, emergent order and dynamic stability). Moreover, conflicting feedback can be experienced simultaneously, sufficient to replicate dynamic equilibrium within a physically or chemically **closed system** (with those factors fostering chaos being countered by those aiding predictability, in the manner of so many forward and back reactions).

Beginning with impalpable prenatal differences, the mind cycle, multi-dimensional deviance and **social learning** serve to amplify humanity's weakness for unsound thinking. Whereas the hypothetico-deductive cycle, received learning and coevolutionary dynamic serve to consolidate **sound-mindedness**, whilst forging stable communities. Consequently, **pathological asymmetries** force a review of *Hominini* – that longstanding taxonomic tribe having become a crucible of radiating possibilities, many at odds with nature itself. In effect, the term **acute nonlinear event** better describes sudden or unexpected changes in otherwise stable **communities** (changes which could be constructive or destructive, internalized or externalized, competitive or competing). That is to say, much hinges upon the rationale of the people themselves – with some aiding predictability and others fostering chaos.

- Cold logic

The Centre-ground of politics is conspicuous by the fact that it pro-actively separates, fragments and denies power without recourse to **subnational**, **national** or **supranational** interventions. The supranational application of telepathically-induced effects enables mainstream agents to fall neatly under the spell of deterministic top-down nucleomotive and electromotive forces – and in a manner reminiscent of **the rule** (namely, "*favour rewarding; use punishment sparingly, and only to define the limits*"). Punishment being an exclusively mainstream activity, heavily-scrutinized by **balanced minds** (minds tasked with rewarding legitimate forms of social control and honourable legal practices). At the heart of personal freedom lies cold logic, whereby an honest appraisal of that which is **biologically-produced** aids the imposition of **socially-constructed** remedies.

As regards the imposition of radiofrequency effects, far too many individuals do what they believe they can '*get away with*' (whether on the roads, in their professional lives or whilst administering the law). Needless to say, every highway betrays that ill-disciplined mentality, with many flouting textbook answers. Those impatiently breaching the stipulated limits effectively barring themselves from applying neuro-cognitive effects. That's because there *are* textbook ways of applying telepathically-induced effects – ways which must never be compromised. To that end, cold logic works in harmony with nature to redistribute power, energy and influence and eschew its undue concentration – beginning, in the first instance, with a cold-blooded appraisal of mainstream agents.

The above rule, which denotes balanced-mindedness, comes originally from childcare, placing something as biologically important as nurture at the heart of this book's professed global values. As we've seen, an agent's cyclical mind is uneasily sandwiched between the biologically-produced genetic substructure and the socially-constructed sociopolitical superstructure. Cold logic affords the substructure and superstructure the freedom to stabilize around a balanced-minded norm, to progress towards an enlightened set-point or to deviate wildly towards unsustainable competition. All the while, **mainstream human agency** is to be pragmatically evaluated, using suitably abstracted data, so as to facilitate balanced-minded interventions. Those living in and loving darkness rendering that dynamic ever more efficient – and hence ever more organic.

Ch6: **Hating the light**

• Cognitive bias

Cognitive bias euphemistically denotes humankind's longstanding predilection for partiality, prejudice and predisposition. The list is endless, but includes belief bias (judging an argument's strength on the believability of its conclusions), selective perception (observations distorted by one's expectations), focalism (an over reliance on initial information), availability cascade (reinforcing a belief through public repetition), group polarization (adopting a more extreme position when part of a team), clustering illusion (overstating the importance of patterns within large data sets), group attribution error (perceiving an affiliate as representative of an entire group), experimenter bias (discarding unhelpful data) and groupthink (when a group ignores essential information, due to consensus).

What makes this congenital weakness so alarming is that favouritism, intolerance and chauvinism translate into **nationalism**, **racism** and **sexism** – with **sound minds** proving unbalanced, **logical minds** proving unsound and **biased minds** proving illogical. More commanding are **balanced minds**, which are resolutely anti-nationalist, anti-racist and anti-sexist. It just so happens that faith spans all of the above rationales. For example, religion once *soundly* served as an insurance against infirmity and loss, the devotee offering their time and devotion in return for help and support in times of crisis. If, by some chance, religion took it upon itself to promote charity, forgiveness and compassion, it would also appear *balanced*. However, religion's doctrinaire insistence on *bias* and *logic* has, if anything, sown the seeds of **misogyny**, prejudice and conceit.

Secular **humanism**, for its part, believes that ethics, science and justice are all we need to build and sustain a better world. In reality, they aren't, in-as-much as *Homo sapiens* is naturally corrupting of the same. That is to say, ethics, science and justice are all **social structures**, whereas human beings are agents. Begging the question, do you believe in the **structures**, or in the agents? Humanists fallaciously claim to have confidence in the agents, whilst actually revering the structures – structures which are foolishly subverted by humans! Thus, **humanist bias** (amounting to unrealistic expectations, born of confusing agents with structures, and *vice versa*) leads, if anything, to a perpetuation of social problems.

Basically, humanism and religion both market unrepresentative models – that is, the former's delusional model of humanity versus the latter's fraudulent depiction of the cosmos. The truth is, humanists gamble on scientists applying their knowledge in a balanced-minded manner, even as *Homo sapiens* worships an expanding list of cognitive biases. In fairness, should mathematics prove ineffectual, we'll all end up facing evolution in the raw, with both agents and structures appearing flawed. The reason this book perseveres with calculus, both linguistic and arithmetical, is because of its '*evangelical faith*' in pragmatism and abstraction. However, it would aid comprehensibility enormously if we evaluated this destabilizing **hominin** presence, so as to **subordinate** all but the most sound and balanced **genes** and **memes**.

• False consciousness

Autocratic regimes concentrate power, sufficient to control **declarative knowledge** (that is, facts, figures and statistics) and **procedural knowledge** (for example, electronic know-how, mathematical formulas and industrial processes).[10] Given that human reasoning rests upon these two indispensable pillars, one can see how a citizen might end up weakened by their ideological manipulation. In fact, all states are strategically motivated to control declarative and procedural information, leaving citizens at the mercy of these ideologically-driven distortions. Distortions which breed **false consciousness**, whereby a spurious picture of society, and the issues affecting the same, is falsely presented by those in power, ostensibly to maintain the prevailing class structure and/or political *status quo*.

Bottom-up social action, whereby that *status quo* is questioned, is frequently motivated by **anti-exploitation bias** (a passionately-held belief that inequality, oppression and misinformation are pernicious and widespread), whilst **top-down** ruling class ideology is driven by **anti-insurgency bias** (amounting to the conviction that those denied commanding strategic intelligence, and who are prone to unsafe levels of cognitive bias, cannot be guaranteed to enter into self-disciplined lobbying, or even comprehend the true nature of the concerns). Where the political middle-ground successfully wins through, however, the **debiasing** impact of coevolution serves to counter this top-down weakness for despotism and bottom-up predilection for disorder.

The manner in which a pluralist society oscillates around a political norm or set-point – due, in large part, to anti-exploitation bias and anti-insurgency bias behaving in an antagonistic manner – can only be fully appraised at the supranational level. In the absence of balanced-minded mainstream agents to remotely advantage, hegemonic democracies which are hostage to mounting distrust are likely to behave in an unstable manner, making anarchy and autocracy more likely. As the state's primary source of power is its economy, the **economic infrastructure** ranks as the 'Achilles heel' of rogue, failed and pariah states – being both the means to bring them to their knees and the key to their swift rehabilitation.

In the case of despotism, the state may be subject to sanctions, investment withdrawn and fixed assets destroyed. With anarchy, however, the situation on the ground may be so acephalous (i.e. *akephalos*: headless) that there's no economic infrastructure remaining, and no one in a position to make declarative or procedural use of the same even if there were. Judged from the perspective of the political scientist, anarchy proves the case for structuralism and autocracy the case for social action. That is to say, we all agree that the people should be free – what's less clear, is how the sociopolitical superstructure ought to be configured? Seen through the eyes of the **Western polyarchies**, what we require is a construct which isn't afraid to redistribute power, capital and influence – in other words, what we require is **neocentrism**.

• Neocentric stratification

Imagine a society stratified according to the adjectives **biased**, **logical**, **sound** and **balanced**. **Neocentric stratification** just happens to be the most commendable of twenty-four possible variations of this **adjectival social stratification** (see Fig. 11). What makes this variation so creditable, is that biased and logical are subordinate to sound and balanced. Neocentric stratification underscores the attributes needed to apply telepathically-induced effects responsibly – making it the must-have inequality, in our age of growing **extrasensory** awareness. In those instances where there are no subjects in a particular category, that category will appear at the bottom of the hierarchy, such that neocentric stratification could still arise in a group comprising only balanced and sound individuals, provided that the balanced individuals occupy the more commanding positions.

In truth, neocentric stratification is a normative ideal, which is unlikely to be achieved one-hundred percent at the macrosociological scale. Nevertheless, it stands as an estimable sociological hierarchy – but not a hierarchy which mainstream agents would ever be trusted to flawlessly achieve and maintain. All of which begs the question, could a less than perfect **neocentric state** safeguard people's fundamental rights and freedoms, in areas as divisive as remote manipulation, mental telepathy and the superimposition of persons? Conceivably, having arrived at a significant majority of balanced-minded people at the top, those applying radiofrequency effects could impose the ideal by remote means. Such that only well-balanced individuals would find themselves benefitting from neuro-cognitive effects.

Figure 11: Neocentric stratification

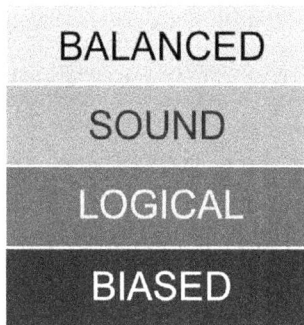

BALANCED

SOUND

LOGICAL

BIASED

The optimal form of adjectival social stratification is neocentric stratification, whereby responsibility for telepathically-induced effects rests with balanced-minded individuals, keenly supported by sound-minded technocrats.

Beyond those stratifications which could be made to work, we meet with **adjectival stratification** and **pathological stratification** (the former denoting a balanced or sound presence at the top, albeit undermined by unsound factions, and the latter signifying a biased or logical presence at the top, sufficient to render it rogue). Adjectival and pathological stratification are unavoidable hazards, arising out of highly-charged national and subnational politics. Which contrasts with stable supranational governance, in which an unchallengeable global construct advantages competitive mainstream democracies. Of course, **plebiscites**, such as referenda, daringly provide for biased minds deposing balanced ones, and for unsound political agendas unseating much sounder alternatives, such is the nature of cold logic.

Accordingly, one might postpone holding a plebiscite, or referendum, until the overarching global framework is securely in place, thereby guaranteeing that there's a safety-net of negative-feedback, should events take a dramatic turn for the worse. By worse, we mean unconscionable social inequalities arising out of nationalism, racism and sexism – such bias and logic leading to **ascription**, whereby groups and individuals are placed in a lower caste, position of servitude, or otherwise handed a politically vulnerable status. Many of society's gravest injustices derive from these ascriptive practices, and eliminating the same must be the professed goal of the political middle-ground. Needless to say, those who do ascribe are to be categorized as clinically unsound, and afforded a subordinate standing.

- Conceptualizing concepts

The term **meme** (i.e. *mimema*: something imitated) was first coined by evolutionary biologist Richard Dawkins (1941–), and denotes an aspect of semiotics or human behaviour which is passed-down in a non-genetic manner. In a hyponymic sense, a meme is a **concept** – or else, behaviour we can conceptualize. Accordingly, thanks to our 'Chomskyan' intellect, our species is able to 'inherit' such notions as **hermeneutics** (the art of interpretation), **miscegenation** (a phenotypic mixing of the races) and the **decomposition of capital** (expanding private ownership, amid a redistribution of wealth). For its part, this book promulgates new memes, through the expedient of neologizing, in order to articulate how biased, logical, sound and balanced become suffused with increasingly sagacious observation – and in order to explain the **cognitive**, **nervous** and **endocrinal** processes responsible for the same.

The least empirically disciplined and most crudely rationalist mindset of them all is biased. Beyond that we encounter logical analysis, incorporating sensory information in an ill-disciplined way. More commanding still is sound, in which experimental scrutiny reaches irrefutable heights. Finally, we arrive at balanced, in which a broader, or more objective awareness, informs human affairs and applied ethics. Constitutionally-speaking, the promotion of social mobility within an **open system** makes neocentric stratification possible. In contrast, **closed systems**, with reduced social mobility, cement religious tyranny, racial tension, political violence and sexual inequality. Therefore, disseminating sound and balanced genes and memes serves as an antagonist to tyranny, tension, violence and inequality.

Cultural relativism maintains that mainstream knowledge, both declarative and procedural, is, without exception, religiously and politically biased (leading to claims of mass manipulation and false consciousness). However, neocentrism would never be so culturally relativistic as to suggest that head-chopping, hand-chopping and the codification of misogyny are mere matters of interpretation. If anything, neocentrism is as debiasing as faith is **biasing**, making it the ideal counterpoint to religious violence. Politically, the **Right-wing**, on hearing that crime is linked to poverty and abuse to obscenity, often races to **austerity** and lax **censorship** as a prelude to scapegoating and **prestige extremism**.[11] The **Left-wing**, conversely, might appear as profligate as it is censorial, such that criminal and abusive tendencies find themselves obscured, as **pecuniary extremism** takes hold.

Disciplined observation enables the political middle-ground to repudiate socially-constructed nonsense, be it pecuniary or prestige in nature. Consequently, a constructive meme is, by definition, one which can be tested by experiment and/or observation – and, if found wanting, abandoned in a manner which discourages its non-genetic transmission. Notwithstanding that third-party refutation may be ideologically-motivated, and its promulgation disrupted for partisan reasons – a case of unsound genes sabotaging the soundest of memes. Somewhat heuristically, hegemonic democracy's top-down '*high-culture*' vies with its bottom-up '*low culture*' to produce a highly-charged milieu, one in which society's deepest and most pernicious stresses get gratuitously amplified – shockingly betraying, by means of some communal **passion**, what humankind really ought to be imitating.

- Pecuniary and prestige extremism

The **primary labour market** (which is noted for its job security, healthy remuneration and career prospects) comprises, amongst other occupations, scientists, engineers and civil servants, many possessing sound or balanced personalities. The **secondary labour market** (which is notorious for its job insecurity, low wages, zero hour contracts and limited prospects) comprises, amongst other job types, many logically-minded administrators, factory workers and retail assistants. Left-leaning politicians see the primary labour market as an obvious target for **progressive taxation**, whereby public finances are boosted without imposing egregiously upon the poorest in society. The secondary labour market, for its part, often

Figure 12: Pecuniary and prestige extremism

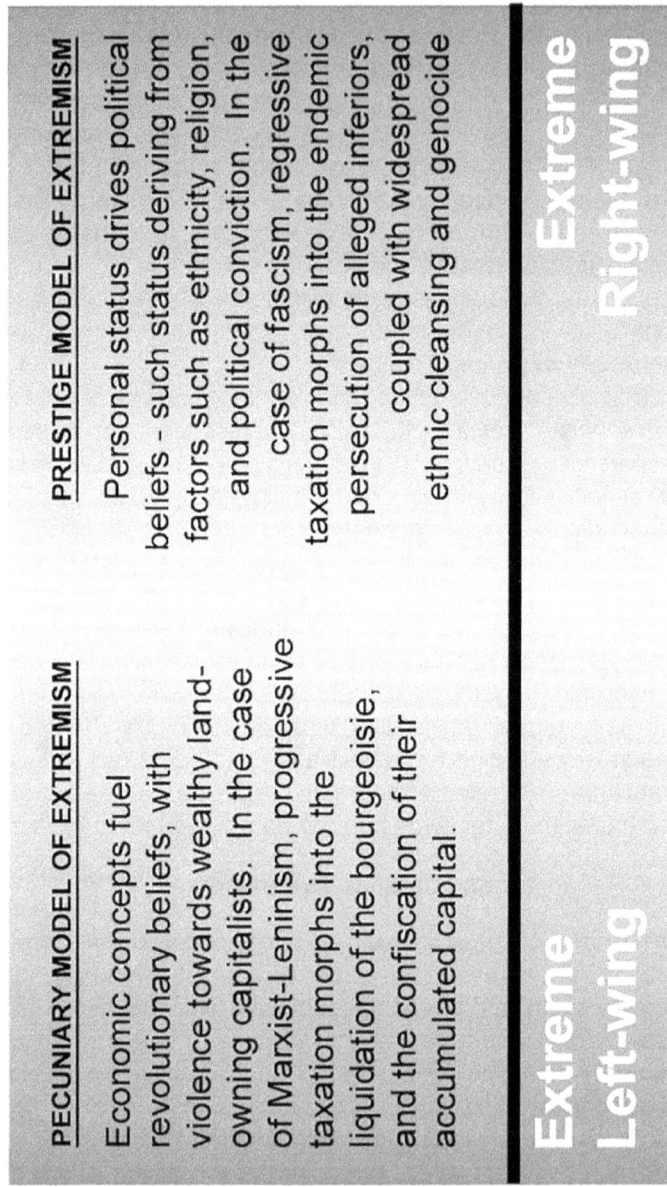

PECUNIARY MODEL OF EXTREMISM

Economic concepts fuel revolutionary beliefs, with violence towards wealthy land-owning capitalists. In the case of Marxist-Leninism, progressive taxation morphs into the liquidation of the bourgeoisie, and the confiscation of their accumulated capital.

PRESTIGE MODEL OF EXTREMISM

Personal status drives political beliefs - such status deriving from factors such as ethnicity, religion, and political conviction. In the case of fascism, regressive taxation morphs into the endemic persecution of alleged inferiors, coupled with widespread ethnic cleansing and genocide

Extreme Left-wing

Extreme Right-wing

falls prey to **regressive taxation**, whereby Right-leaning politicians introduce taxes which fall disproportionately upon the poor.

In the case of the extreme **Left**, e.g. **Marxist-Leninism**, progressive taxation morphs logically into the liquidation of the capital-owning upper and middle classes, together with the confiscation of their accumulated wealth. In the case of the extreme **Right**, e.g. **fascism**, regressive taxation morphs logically into the wholesale persecution of alleged inferiors, together with their premeditated murder (see Fig. 12). Such pathological stratification involves logically-minded functionaries supporting and maintaining biased **patriarchies**. Balanced-minded individuals (such as those charged with applying radiofrequency effects) have as their **primary responsibility** the promotion of political freedom within an open system possessing enhanced social mobility. However, their **secondary responsibility** is to the decomposition of extremist power, which is achieved indirectly through mainstream human agency.

Left-wing extremists conform to a **pecuniary model of extremism**, in which economic concepts fuel revolutionary beliefs; the greatest violence being aimed at wealthy, often land-owning, **capitalists**. **Right-wing extremists**, conversely, conform to a **prestige model of extremism**, in which personal standing is of prime importance; such status deriving from ethnicity, religion and/ or political conviction. **Islamic fundamentalism**, for example, falls under the prestige model of extremism, being on a par with fascism in its use of ascription, bigotry and violence. **Left-wing incitement** to commit various infractions, and **Right-wing scapegoating** in pursuit of a hardline appearance, both serve to escalate pecuniary and prestige wrongdoing.[12] In the case of **Islam**, that scapegoating often takes the form of macabre public spectacle, typically involving the perverse use of amputations.

Extremists inevitably seek to exploit power. Which, in the case of the Left-wing, means reallocating unproductive capital by force. The Right-wing, for its part, disingenuously encourages criminality and abuse (as with Islam's sexual exploitation of **underage brides**, or the police service's closing of ranks in respect of unexplained deaths in custody). Clearly, the Right-wing glories in its pretense of decorum, whilst debasing itself privately. As for the Left-wing, it's quite happy to debase itself publicly, provided that financial inequalities are ruthlessly eliminated. Providentially, neuro-cognitive powers enable balanced minds to remotely 'observe' these illegalities – and not simply the scarcely-concealed

61

desperation of Left-wing agitators, but also the laughable charade of professed Right-wing values.

- Was Hitler logical?

British-born professor Michael Mann (1942–) contends that **social power** takes two principal forms: firstly, **distributional power**, whereby an individual gets others to carry out their will; and, secondly, **collective power**, whereby a group gets others to carry out their combined will.[13] Additionally, **extensive power** denotes having power over many people; and, **intensive power**, a high degree of compliance from one's subordinates. As for submission, that implies **authoritative power**, whereby less qualified subordinates consent to obey. **Diffuse power**, on the other hand, implies compliance without recourse to express commands. Furthermore, Mann differs from Marxists, in-so-far as he maintains that, in addition to economic power, there are also political, military and ideological sources – and that these sources act more or less independently.[14]

Mann further surmises that **globalization** subverts the state's powers, rendering concepts such as 'society', and even 'referenda', comparative abstractions (the macrosociological and microsociological appearing to mirror the conceptual and corporeal, as regards their relative importance). That said, the **UK**'s recent **EU referendum** – in which the British people were invited to choose between being '*in, to opt-out*' or being '*out, to opt-in*', as regards European Union membership – proves that the English, at least, remain awkwardly attached to their 'foggy' geopolitical apparition.[15] Of course, English nationalists would much prefer that their powers weren't being eroded by post-imperial geopolitical trends – in which case, they ought to jolly well stratify neocentrically, sufficient to command a strategic role within **NATO** (the only alternative, as the Americans can attest, being world domination and the subjugation of the global economy).

The devastating irony, of course, is that political, military and ideological powers all become contingent upon economic power, through successive globalization. The first signs that such a codependence might sow the seeds of chaos, criticality and catastrophe came with the **Wall Street Crash**, in 1929. Logically, to break-free of this nascent globalization, Adolf Hitler (1889-1945), who'd acceded to the position of German Chancellor in 1933, needed to subjugate the entire world economy. Thus, the Nazi Party began its wretched path to global domination, commencing with a clear-cut prestige model, demanding the pathological stratification of German society along racial, religious and ideological lines. Unbeknownst to Hitler, however, his unsound aspirations would eventually render the resultant negative-feedback enviably efficient.

The Nazi 'brownshirts', and later Wafen SS, were both emblematic of Hitler's intensive power base. Arguably, he sought to liquidate **Zionism** *en masse* (by which we mean the commitment to establishing a Jewish homeland within the British mandate of Palestine) in order to win favour with the Arabs, sufficient to access Middle-Eastern oil. Answerable to Hitler, at that time, were logical minds, capable of deducing the strategic 'benefits' of the fuhrer's **anti-Semitic policies**, including the Holocaust itself. With dissenting voices silenced, foreign policy fell into the hands of biased anti-Semites, gambling on winning-over **Arab nationalists**. A strategy which, following the Stalingrad debacle, became ever more murderous – leading, by the war's end, to the blanket bombing of German cities, as western strategists self-interrogated over where to draw the line.

In the event, World War Two's adjectivally stratified victors found themselves besieged on all sides by bias and logic, bringing many postwar constructs into doubt. Zionists and Arab Nationalists, for example, became directly pitted against one another, through the establishment and expansion of the state of **Israel**. What an invisible hand can do, given the corrosive impact of biased and logical factions, is selectively advantage those Americans, Germans, British, Israelis, Palestinians, etc, who appear sound and balanced. As for those Arabs, Jews, Christians, Brexiteers, etc, persevering with unsound agendas, the growing demand for permanent political solutions may well see them meet with a series of acute nonlinear events. Catastrophic for them, no doubt – but, crucially, a means of avoiding never-ending chaos. After all, consensus is nothing short of incendiary, in a nation harbouring an unsound majority.

PART III

Neocentrism

Chapters 7-9

Ch7: Separation, fragmentation and denial

•   Entropy, ecology, evolution and emancipation

The UK, USA and EU don't exist as discrete entities, their borders far too blurred by foreign ownership, cultural diffusion and climatological exchanges. By the same token, atoms, organs, organisms and communities don't exist as discrete entities either. In other words, boundaries are convenient abstractions, sustained by means of mathematical pragmatism. Physicists, biologists and sociologists might all assert that there are only levels of organization, custom-made to describe the separation, fragmentation and denial of mass, arising from power, energy and influence. That is to say, should the United Kingdom ever succeed in achieving unqualified independence, it will have the appearance of chronic organ failure – such is the nature of our mutual, not to say mortal, codependence.

Figure 13: **Cruciform calculus**

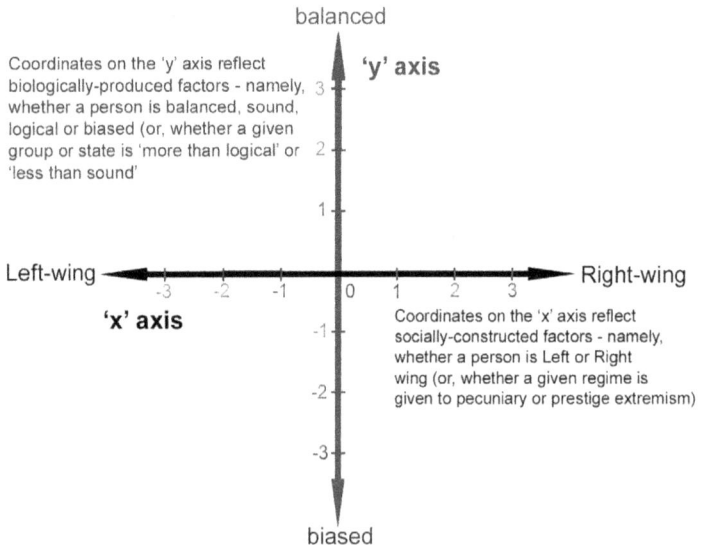

balanced

Coordinates on the 'y' axis reflect biologically-produced factors - namely, whether a person is balanced, sound, logical or biased (or, whether a given group or state is 'more than logical' or 'less than sound'

'y' axis

3

2

1

Left-wing ←———————————————→ Right-wing

-3   -2   -1   0   1   2   3

'x' axis

-1

Coordinates on the 'x' axis reflect socially-constructed factors - namely, whether a person is Left or Right wing (or, whether a given regime is given to pecuniary or prestige extremism)

-2

-3

biased

With that in mind, neocentrism blurs **sovereignty**, much to the chagrin of those endeavouring to stratify pathologically. The overarching superstructure (that is, growing international collaboration at the supranational scale, amid the pretence of autonomous national and subnational politics) is best illustrated by a **cruciform model** (see Fig. 13). With arithmetical axes enumerated to reflect both biologically-produced and socially-constructed factors, its Cartesian coordinates reflect specific levels of organization. Contrary to appearances, this coevolutionary dynamic, or **duel inheritance** mechanism, is actually a seamless cycle, which can only be modelled by ignoring many of the details. Those who begin with sharply-defined concepts, do so because they're ill-acquainted with dynamic equilibrium as it pertains to healthy open systems.

As we've seen, realists, Platonists and neo-Platonists view mathematical objects as objectively real, whereas anti-realists, conceptualists and constructivists see them as unquestionably subjective. The latter might even go so far as to suggest that mathematical objects, sovereign boundaries and God are all equally subjective – and in spite of passionate calls to the contrary from those serving science, politics and religion. The cruciform riposte, of course, is to think of sound and balanced individuals as objectively capturing the truth, by pragmatically constructing abstract objects (whereas biased and logical individuals accept as true only such objects as they've subjectively envisioned). Accordingly, the cruciform model serves as a precursor to capturing the truth – not least, by discarding a wealth of factual detail.

Paradox, in the sociological sense, is evidence of biased and logical deliberation, with cruciform analysis, as outlined above, routing-out any likelihood of the same. In the law, for example, the malign use of telepathically-induced effects gives rise to the **mitre paradox** – namely: "*the defence of having been abused is admissible, in the case of an alleged wrongdoing, provided that the abuse can be proven. However, if the abuse is provable, why transgress, why not simply prove that one was being abused?*" This legal paradox presents a defendant with a distressing Catch-22 scenario, in-so-far as they may find themselves subject to alarming levels of remote interference, none of which can be proven. One must assume that the law expects people to suffer years of sustained radiological abuse, with little or nothing said in their defence.

- Decomposition of power

Globalization is constructive because it subdues mainstream powers. The imperialism which preceded it was the very antithesis of globalization, in-so-far as the world economy was subjugated by competing European interests. Such imperialism led inexorably to the Seven Years War (1756–63) – which, according to the

British historian Niall Ferguson (1964–), "*was the nearest thing the eighteenth century had to a world war*" – plus two subsequent global conflagrations.[16] If imperialism is defined as a **concentration of power**, by way of colonialism, then globalization represents a decolonializing **decomposition of power**. That decomposition encompassing economic, political, military and ideological powers (as opposed to the decomposition of capital, whereby economic power alone is disbursed). As well as sparking world wars, imperialism was unwaveringly biased and logical – as evidenced by Britain's Atlantic slave trade.

Britain's abolition of the Atlantic slave trade, in 1807, denotes an infusion of balanced-minded political reform, stemming from a sound dissemination of knowledge – much of it attributable to the distributional influence of William Wilberforce MP (1759-1833). As such, it was the growing separation of domestic political powers which kick-started the dissolution of this unsound racist trade, with the emancipation of British-held slaves following in 1833. However, in the eyes of social reformers, like Karl Marx (1818–83), that still left the colonialist's very own **proletariat** at the mercy of class bias and capitalist logic. Nevertheless, the political landscape was beginning to provide for the distributional impact of conscientious lobbyists, such lobbying occurring within newly **disaggregated states**. In other words, states ripe for reform.

In the same way that capital decomposed, thanks to grassroots embourgeoisement and the class-system's eventual breakdown, British imperial power decomposed, thanks to an impassioned 'rights of man' debate at home. Clearly, the sociopolitical superstructure, needed to make meaningful use of these conscientious exchanges, was very much in place. However, the demise of imperial power doesn't come with a guarantee that murderously-biased groups and logically-unsound individuals won't fill the resultant void, or that the infrastructure of progressive change will somehow spontaneously appear. Thus, when the British withdrew from India, in 1947, following Mohandas K Gandhi's (1869-1948) campaign of non-violent non-cooperation, it unleashed a wave of ascription-fueled depravity between Hindus and Muslims.

Indian independence, as it turned out, wasn't some clear-cut decomposition of power, arrived at within an open-minded disaggregated structure – but rather, the narrow-minded pursuit of unsound religious interests, gratuitously amplified by assumed spiritual differences (resulting in the formation of Pakistan, India and Bangladesh). Without doubt, any domestic stability which has subsequently arisen, scarcely compensates for the alarming regional instability which has followed, when one calculates-in a nuclear weapons capability. Proving, beyond question, that even when newly-formed nations survive morally repugnant forms of **ascriptive cleansing**, the regional picture which follows is often psychologically disturbed – bequeathing the 'as-yet-unborn' a monumental trigger for future thermonuclear conflict.

Gandhi was ostensibly correct, had the '*ungodly*' masses acted in accordance with his wishes, a less dangerous world would've materialized. Instead, biologically-produced bias and logic met with socially-constructed sophistry and prejudice, sufficient for '*evil*' to manifest itself. As luck would have it, **Western science** has the means to counter that diabolical synergy, aided by globalization. Globalization makes the sound and balanced decomposition of capital easier to initiate, and the biased and logical acquisition of political, military and ideological powers harder to sustain. What's more, globalization is the ideal starting point when faced with regional disorder, economic imperatives often transcending ascribed differences. In other words, provided the state possesses a disaggregated political structure, multi-ethnic military and balanced ideology, the resultant economic community ought to be enviably stable.

- Defragmentation, reunification and assent

Federal unions, bound by the free movement of goods, services and people, aren't imperial powers, e.g. the **United Kingdom** (UK), **United States of America** (USA) and **European Union** (EU). Of these, the EU stands head and shoulders above the other two, being the only one to have established itself solely through democracy and consensus, albeit on terrain blasted-flat by warring – the UK and US both being products of quasi-imperialism involving considerable amounts of violence and coercion, much of it directed against their own populations. To survive, however, the US and UK have had to recast themselves in the mantle of winsome federal unions. In large part, due to twentieth century imperial decay – a decomposition of power so great it resulted in the collapse of the Ottoman Empire, the sun finally setting over the British Raj, French colonialism foundering, the Axis powers capitulating and the USSR theatrically imploding.

If Italy could be said to have exported the **fascist** beliefs of Benito Mussolini (1883-1945) and Great Britain the **communist** dogma of Karl Marx (1818-83), then America might legitimately claim to have marketed George Washington's (1732–99) incipient **federalism**. Today, the political Centre tends towards federalism; and in spite of a colourful history, much of it peppered with pecuniary and prestige excesses. Today, the UK has grudgingly embraced federalist principles, as evidenced by its **devolution** of powers to regional parliaments and assemblies – although the Northern Ireland Assembly's current suspension does smack of quasi-imperialism, as does Westminster's continued cowing of the Celtic fringe in relation to **Brexit**. However, once anti-European English nationalists wake-up and realize that federalist healing is the future, and that the baneful use of direct rule is hypocritical, that anachronistic impasse will be broken.

Figure 14: Concentration and separation of power

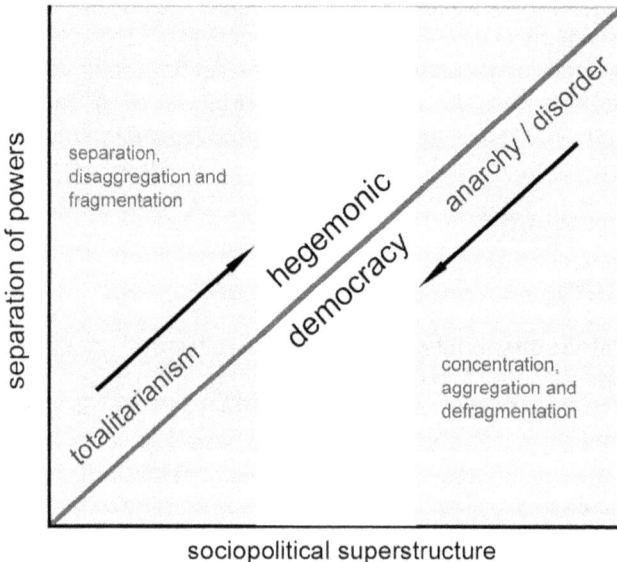

sociopolitical superstructure

Constructive pluralism, otherwise hegemonic democracy, demands
that power be either concentrated or separated, as the case demands
(with mainstream responsibility, as regards foreign intervention,
resting, in principle, with the political middle-ground).

As we've seen, kingdoms, colonies and cantons don't always neatly disintegrate into open-minded disaggregated structures, hell-bent on democratization, federalization and globalization. In fact, in the event of acephalous disorder one would need to **concentrate**, **aggregate** and **defragment** those aspects which unite to produce hegemonic democracy, whilst differentially discriminating against the unsound alternatives (see Fig. 14).

70

America, of course, would argue that it has forcefully concentrated, aggregated and defragmented those aspects of itself needed to build and sustain a fully working democracy. However, translating that procedural and declarative endeavour overseas is fraught with difficulties, due to many foreign states possessing unsound majorities. That is to say, the US surmounts, or continues to surmount, its very own politically imperious consensus.

As recently as 1959, Alaska and Hawaii acceded to the US, becoming the 49th and 50th states of America respectively. As for the European Union, it has similarly extended its borders, most conspicuously in 2004, with the accession of countries formerly allied to the Soviet bloc. Whether **Brexiteers** admit it, or not, federalism remains an important post-colonial trend. Indeed, the UK's survival depends upon it advancing federalist principles and breaking new ground in respect of devolution. Therefore, a much closer cooperation between the UK, US and EU is conceivable, as they all speak the language of democratization, federalization and globalization. A language which enables them to exert a humanitarian influence well beyond their respective borders – whilst all-the-while rendering themselves politically, militarily and ideologically vulnerable. That creeping vulnerability being felt most acutely by those who are palpably 'less than sound'.

- Germany's tides of power

Following their defeat at the hands of Germanic rebel leader Arminius (c.18BC–AD19), in AD9, the Romans candidly chronicled their own imperial shortcomings. The subsequent internecine warring between those triumphant Germanic tribes doing more to divide those bound by common language and custom, than unite them. Much later, the Frankish, Saxon and Hohenstaufen empires would all similarly crumble – leading, in turn, to a more sustained disintegration of bellicose 'German power', arising out of escalating religious differences. What followed were the Wars of Religion (1618–48), which left the Germanic heart of Europe fragmented into copious territorial units. Territories which, in the north, began to coalesce into an overtly militaristic Prussian state. Prussia being looked upon as the logical manifestation of Germany's First **Reich** (that is, prehistory-1871).

The French Revolution, beginning in 1789, fuelled mounting German nationalism, what with Prussian soldiers standing shoulder-to-shoulder with the British at the Battle of Waterloo, in 1815, thereby helping to administer the *coup-des-gras* which finally ended France's Napoleonic imperialism. The British, for their part, were entering an age of reform, sufficient to avoid a similar revolutionary meltdown. In time, the Westminster model, which those reforms wrought, would commit Britain to governing from as close to the political Centre as circumstances would permit. Had this English enlightenment proceeded more swiftly, her imperial resources might have been used to decide the outcome of Prussia's 'Seven-Weeks War' with Austria (1866) and the Franco-Prussian War (1870–71). Thereby countering this fledgling German nationalism.

In the event, failing to oppose Prussian belligerence, personified by Otto von Bismarck (1815–98), gave confidence to truculent German patriots. Driven by their unsound military logic, Germany quickly appropriated the mantle of a Second Reich (1871-1918). Regrettably, a sound **political meridian** had yet to fully establish itself within the British Isles. Thus, an acquisitive German Federation ascended out of the unstable tribalism of the First Reich, whilst Britain's liberal reformers looked on. From its very inception, Germany's political superstructure, comprising a Bundesrat (Federal Council) and Reichstag (Imperial Assembly), was ridden with Prussian bias and calculating military logic. That is to say, had a progressive federalist model been pursued from the outset, a successful Centrist state might well have emerged.

With devastating irony, Germany advanced its imperial ambitions just as colonialism was foundering. Bismarck's Reinsurance Treaty, arrived at secretly with **Russia** in 1887, epitomizes the tensions surfacing at that time, as Germany sought to avoid Russo-German, Franco-German and Austro-Russian wars. A move which, depending on your interpretation, maintained the European peace for a generation or cemented tribal differences. In the event, the healthy mind paradox interceded, self-aggrandizing Germany's ruling class, who provoked, what they took to be, Russo-German and Franco-German wars. What they met with, of course, was the wholesale annihilation of imperialist fantasy by virtue of them igniting World War One (1914–18). Thus, Germany's Second Reich ended in humiliation and defeat.

However, democratization, federalization and globalization still beckoned; beginning with a revised German constitution, demanding that the Reichstag be elected every four years, via proportional representation, and which sensibly incorporated universal suffrage and votes for women. Therefore, thanks to US President Woodrow Wilson's (1856-1924) Fourteen Point Plan, a fully-fledged German democracy arose out of the carnage of the Second Reich. However, this Third Reich (1918–45) quickly witnessed a sinister change of direction, following the Wall Street Crash, in 1929. Financial codependence had rendered the German economy vulnerable, and in spite of the German people appearing to acquiesce over political, military and ideological weakness. In other words, but for economic collapse the German people may well have renounced imperialism and become altogether more democratic, federal and globalizing.

As it was, the German electorate fell for the prestige fantasy presented to them by Adolf Hitler (1889-1945). Globalization was still in its infancy, which would explain why it erroneously punished compliance, leaving Germans sympathetic to Nazism's brand of murderous colonialism. By the end of the ensuing Second World War (1939–45) the decomposition of Nazi power was so complete, that currency reforms and the division of Germany turned the subsequent **Cold War** (1946–89) into a contest between contrasting economic systems. In the words of English historian Michael Balfour (1908–95), "*the essence of the West German 'miracle' was that at a time when world demand was high, German industry found itself with spare capacity, the right products, and relatively low costs, so that exporting was easy and profitable*".[17]

Balfour argues that wage restraints, surplus labour, weak unions, industrious thinking, zero defence spending and forced modernization all encouraged foreign investment, laying the foundations for the German economic miracle which followed. Today, the unified **Federal Republic of Germany**, which Chancellor Angela Merkel (1954–) currently presides over – albeit with some telling neuro-cognitive tremors – boasts Europe's largest economy. It's been a painful lesson, but Germans have finally grasped that the **global financial system** is coevolving, and has been ever since colonialism's last breath. Nevertheless, these events prove two things: firstly, that globalization is the ideal climate within which to rehabilitate countries; and, secondly, that **unsound factions**, in spite of sowing the seeds of untrammeled chaos, clearly understand the value of economic rehabilitation.

• **The European future**

The Brussel's based European Commission is answerable to every EU citizen via the European Council, comprising the various heads of government, and through the directly-elected European Parliament (indeed, the Parliament can dissolve the Commission, a

Commission proposed by the Council). Therefore, the configuration of the EU is democratic, with 'Brussels' its executive branch. As its perceived deficiencies stem from voter apathy, media cynicism and a lack of engagement, those shortcomings reflect upon the detractors themselves, rather than on the EU. Looking back, Britain joined the **European Economic Community** (EEC) when the global economy was in a state of turmoil, following the dollar crisis (1971). The question then, as now, wasn't whether we should answer to 'Brussels', 'Westminster' or 'Washington', but rather, whether acute nonlinear events were being handled in a biased, logical, sound or balanced fashion.[18]

Like the UK, the EU and USA are committed to governing from as close to the political Centre as practicable, that commitment being achieved through the expedient of sound and balanced thinking (or, failing that, through biased and logical factions, on both the Left and Right, behaving antagonistically). That assurance is important, because neocentrism implies a mechanism which is capable of advantaging the political middle-ground. Thus, stressful **nonlinearities**, acting in the manner of so many *agent provocateurs*, betray which states, or federal unions, warrant a **supraordinate** or **subordinate** status. **Supraordination** serving as an alternative to the now obsolete notion of **superpower status** (with democratization, federalization and globalization lending themselves to supraordination, and totalitarianism, nationalism and imperialism to subordination).

Those who've read '*The Oxford Illustrated History of Britain*' (OUP, 2009) can't help but appreciate the enormous significance of a **single integrated economy** (be it in terms of social progress, regional stability or raised living standards). Indeed, the European Union owes its very existence to that elementary capitalist logic, beginning, as it did, with the European Coal and Steel Community (ECSC). In retrospect, '*the free movement of goods, services and people*' might have been better framed as '*the free movement of goods, capital and labour*', in-so-far as labour is synonymous with profit, whereas people suggest increased public expenditure. So, whilst Brexit has generated an impassioned debate regarding Britain's internal and external borders, the labour-intensive **markets** would much prefer to ignore the same (economic miracles deriving from factors other than national boundaries).

Ideally, a federal union would contain one or more **supraordinate states**, capable of collaborating as a **global governance complex**. But let's not be naïve, even the most democratic state contains

Right-leaning and Left-leaning cronies, whose actions bring into question their political elite's assertions of cognitive, nervous and endocrinal preeminence. For example, to what extent has the murder of black teenager Stephen Lawrence (1974–93) been publicly reflected upon in Great Britain in order to draw attention away from suspicious deaths in police custody or asylum detention? **NEURON**, the global governance complex's neuro-cognitive wing, must hold even supraordinate states to account, if we're to avoid becoming victims of media-led smokescreens.[19] To that end, supraordination presupposes stratification within an open system – a sociological balancing act of such precariousness, that it could just as easily result in subordination.

- Stratified extrapolation

Stratified extrapolation takes the terms 'logical' and 'sound', and supplements them with 'biased' for less than logical, and 'balanced' for more than sound. These terms can then be used to make predictions, based upon their anticipated frequency. Accordingly, in same way that science predicts climatological change and meteorologists forecast the weather, stratified extrapolation estimates how information is likely to be conceptually processed. For example, will a 2.4^{0}C increase in the average global temperature by the end of the 21st century meet with less than logical behaviour? And, to what extent will anti-insurgency bias and anti-exploitation bias dominate US domestic affairs? Clearly, it's not simply about anticipating the fate of the proverbial ship, it's also about predicting the actions of the hypothetical passengers and crew. If accurately appraised, it may be possible to forecast their behaviour before catastrophe strikes.

Revealingly, a fascist and a communist were once stood discussing the Titanic disaster. The fascist argued that there was a surfeit of lifeboats, given that steerage class needed to be kept in bondage on the lower decks. The communist, however, retorted that had steerage class been freed from their bondage, they could've torn-out the ship's useless aristocratic adornments and crafted the means of their collective survival. Clearly, even if the iceberg can't be avoided, neocentric stratification affords the passengers and crew some prospect of survival, with many on board petitioning for 'a lurch to the Left'. Indeed, it's much easier to stratify in that manner, than anticipate every eventuality. Besides, how else could unsound factions surmount calamities directly proportional to their lack of reasoning, without sound and balanced interventions?

Democracies, oscillating between Left and Right, might burden themselves with fatal misjudgments, and even begrudge others being shielded from their lacklustre critical thinking skills (for instance, when ideological antagonists waste valuable time clashing over the dictionary definition of **anti-Semitism**, whilst Rohingya children are having their throats slit). That said, with the global financial system currently coevolving to reward sound and balanced behaviours, those genuinely concerned with humanitarian and environmental causes will find that in the future they have the power, capital and intelligence needed to act decisively – the need for such interventions, on the part of the political middle-ground, having been shrewdly prophesized in advance. However, given the range of responsibilities which are likely to be borne by the Centre-ground, it seems prudent to critique the same and extrapolate as to its future.

Ch8: **Political middle-ground**

• L'Esprit des Lois

The financial system is extraordinarily fluid, not only does the value and availability of money fluctuate, so too does the value of the items it purports to represent. Economists have found that when **demand** increases, **supply** rarely keeps abreast of the same, causing prices to rise, triggering inflation. Right-wing **monetarists** logically reduce the availability of money, by means of **austerity**, in order to suppress demand (whilst encouraging savings through **interest rate** rises, breaking the unions in order to curtail wage demands and by cutting welfare spending). Left-wing **Keynesians**, conversely, believe that **economic growth** follows logically from meeting rising demand (which might involve increasing industrial output through financial investment, reducing unemployment by expanding the public sector or introducing **anti-austerity** measures).

Not unreasonably, one might ask whether supply is inevitably deficient, or whether it's sluggish due to avoidable deficiencies. More specifically, is output hampered by prevailing conditions or by the temperament of working age individuals? The Right-wing is inclined towards dispositional attribution and the Left-wing towards situational attribution – that is, each shows signs of ideological bias, with economic woes variously attributed to the citizenry itself or to the wider economic climate. A dichotomy made worse by virtue of the Left-wing and Right-wing both attracting large numbers of biased and logical devotees, whose attributional inaccuracies swell the ranks of the **militant-Left** and **far-Right**. The political middle-ground, on the other hand, eschews unsound attribution, thereby avoiding the political excesses of monetarists and Keynesians.

French political philosopher Baron de Montesquieu (1689-1755) famously penned *De l'esprit des lois* (The Spirit of the Laws, 1748), in which he argued for a separation of executive, legislative and judicial functions. Whilst helpful, it would be prudent to expand upon that longstanding depiction of **Centrist politics**. Building upon the same, an unmanaged economy, subject to equal amounts of **inflation** and **deflation**, maintains stable wages and prices. However, the material expectations of today's electorates has caused pluralist economies to become heavily managed, largely to reduce inflation and **unemployment** – the Right-wing risking unemployment and the Left-wing inflation, with the actions of independent national banks helping to confound that generalization. By definition, a politically neutral national bank manages people's expectations, amid recurrent **boom-bust cycles**, arising out of innovation and obsolescence.

Austerity remains controversial, owing to its impact on **relative poverty** and **absolute poverty** (the former suggesting an absence of broadband connectivity and the latter indicating food-bank reliance). In truth, once austerity has begun to aggravate absolute poverty, the Right-wing has lost the argument regarding its social and economic benefits. Those on the Right-wing, who dismiss rising destitution on dispositional grounds, are prestige diehards, whose actions aggravate grassroots distress. Conversely, Left-wingers, who see only situational grounds for aggrieved **proletarians**, are pecuniary fanatics, prone to weakening the establishment. In other words, those charged with applying telepathically-induced effects must remain ideologically neutral, not unlike a national bank's governor, if they are to adequately defend the political middle-ground.

• Hard-fought freedoms

To label 'Brussels' undemocratic is fallacious, in-so-far as it has placed the needs of EU citizens above those of the European Union itself. Whereas economic reasoning demands **flexible immigration** in support of temporary **stabilization measures**, the EU has actually enabled its citizens to behave in a wholly self-interested manner. Consequently, European migration, which is neither temporary nor flexible, has the potential to confound economic reasoning – notwithstanding, that economic reasoning is a form of capitalist logic, and hence technically unsound. *Ergo*, the EU is correct to place sound and balanced principles above the logical and biased markets. In effect, forcing European nations to draw upon **Keynesian laws** and **monetarist principles** when managing destabilizing, inflexible and enduring demographic changes.

Fortunately, neocentric stratification promotes balanced-minded observers to positions far exceeding mainstream politics. Which is vital, given that mainstream politics is ridden with **political logic** and **political bias** (its politicians presenting valid and invalid arguments respectively). As we've seen, adjectival stratification denotes a democracy which might achieve neocentric stratification, and hence form part of a global governance complex, pathological stratification notwithstanding. That global governance complex, in a manner reminiscent of today's federal unions, would endeavour to relegate the militant-Left and far-Right – effectively pitting logic against logic and bias against bias at the ballot box, whilst soundly encouraging the people to democratize, federalize and globalize.

Without social equality, self-actualization becomes dependent upon inherited wealth and title, rather than upon open-competition and merit. Consequently, **socialism** stresses a reduction in material inequality, together with access to education and advancement. Healthcare, for its part, has become highly politicized, with much said about its appetite for capital and labour. Arguably, if healthcare were in private hands, competing salaries would be used to entice prospective employees, especially in the absence of migrant labour. In other words, private ownership, like a bitterly-fought trade war, would aggravate the inflationary **wage-price cycle**. Making the **'tax and spend'** solution, revered by **socialists**, sound. Thus, a national health service, replete with competitive salaries, is a veritable must-have – with NEURON similarly funded.

As one's heteronnubial co-author wryly observes, immigration, inequalities and icebergs all tease-out the unvarnished truth, as regards society's biologically-produced substructure (not least, through the social constructs they provoke). One could argue that "**exogeny** impacts upon **endogeny**, producing an endogeny which has all the hallmarks of exogeny" (that is, externally stressed agents fabricate social structures comparable to external stresses, with the political middle-ground torn between **centralization** and **decentralization**). However, the solution to unsound capitalist decentralization isn't unsound centrally-planned socialism, but rather, as many have professed, capitalism tempered by socialism – the global financial system looking like an extension of the political middle-ground, rather than an imposition on the same.

• Non-negotiable rules

Codification affords a constitution distributional power, the law a collective power, case law an authoritative power, doctrines an extensive power and maxims an intensive power. From the **European Convention on Human Rights** to the 'best evidence rule', the political middle-ground values non-negotiable codes, statutes and conventions. One way to evaluate human reasoning is to set-out balanced protocols, and then contrast observed behaviour with the same. Being careful, of course, not to bring one's own reasoning into question with **fundamental attribution errors**, whereby too much blame is assigned to another's **disposition**, and too little weight given to the demands of their situation (or **actor-observer bias**, in which another person's disposition is criticized, but similar erring on one's own part would be attributed to one's circumstance).

Throughout the 1960s iconic personalities, such as Martin Luther King Jr (1929–68) and John F Kennedy (1917–63), inspired loyalty. Today, such faith has largely dissipated, with many key voices summarily silenced on the basis of hypocritical mud-slinging, leaving only plain-speaking standard texts in their place. Allied to that undying support for the written word, is the Centre-ground's deep affinity with history – but not the self-glorying patriotic kind, so much as barefaced self-interrogation. Making neocentric historical narratives profoundly attributional, in-so-far as they tease-out the truth regarding disposition, whilst critically appraising the impact of circumstance (not least, the role of radiofrequency effects). Accordingly, key voices may be on the decline, but as one's co-author's presence keenly illustrates, inspirational voices needn't be typically mainstream, with telepathy offering a novel disaggregated dimension to socially-concerned debate.

Strictly-speaking, codification is a social structure, engineered to resist human meddling – with disillusioned anti-humanists resolving to make that construct as resistant as possible. Indeed, just as the electronic configuration can stop a bullet, the global governance complex would force humankind to alter, rather than the biosphere (producing a hominin less heavily-reliant upon treaties). And so, whilst *Homo sapiens* constitutes a logical and biased branch on the tree of life, its replacement, ***Homo aequipondium***, represents a sound and balanced alternative (poised to be competitive, rather than unsustainably competing). **Sapiency** has given us, amongst other things, transatlantic slavery, weapons proliferation, targeted genocide, mass extinctions and climatological change – the hope is that **aequiponimity** will correct that imbalance, affording those who possess it a lasting ecological niche.

Postmodernism implies that the political Centre has achieved international primacy, and is therefore setting the future global agenda – a neocentric agenda which provides for, amongst other things, far greater freedom of expression. However, like so many agendas, the price tag may appear prohibitive. That is to say, to voluntarily submit to progressive taxation, sufficient to fund **cruciform coevolution**, the primary labour market would probably need to be **aequipine**. Unfortunately, many mainstream taxpayers are **sapient**, and dedicated to prolonging **sub-aequipine** activities. The answer, of course, is for those evading their fiscal responsibilities to be categorized as subordinate, and to bear the label *Homo sapiens* – placing them, and possibly their families, in a less commanding position.

- Unproductive capital

Economic interventions are used to steer advanced economies, either through **fiscal policies** (related to **tax revenue** and **government expenditure**) or through **monetary policies** (involving the national bank's management of **interest rates** and **money supply**). Customarily, intervention has sought to raise or reduce **aggregate demand**, because without such measures a fall in demand may result in **overproduction**, and escalating demand, born of supply-side perturbations, in **underproduction**. Problems made worse due to **supply and demand** operating at different speeds, commonly leading to **inflation** or **stagflation** (deflation having become something of a rarity in **The West**, post-World War Two).[20] The Centre-ground subscribes to economic interventions, if only to help supply and demand operate in a suitably rewarding manner, via well-balanced fiscal and monetary policies.

To increase productivity, a government may introduce a wealth tax on unproductive capital, thereby encouraging its commercial utilization. Likewise, lowering interest rates may expand an economy, but will decrease the **exchange rate**, causing the currency to depreciate relative to other currencies (a weak currency negatively affecting imports, whilst boosting exports). Here we enter the realms of **currency wars**, in which uncompetitive nations strive logically to compete – their unsound capitalists being compelled by *austerity and decentralization* and their unsound socialists by *centralization and stimulus*. Ironically, the free market, so prized by Right-wingers, all but obliterates state boundaries, whilst socialist planning, so valued by Left-wingers, inadvertently strengthens the same.

Brexit has divided the **Conservative Party**, because its members can't agree on whether to permit the markets to freely erode their beloved sovereignty. Equally, the **Labour Party** can't agree on whether to centralize the British economy, thereby strengthening Britain's national boundaries. Either way, globalization cleverly frustrates those not occupying the political middle-ground, with some on the Right and Left appearing chagrined with both capitalism and socialism. Looking ahead, a global governance complex would respect the free market's appetite for worldwide integration, whilst harbouring unambiguous socialist aspirations. Moreover, its stratified extrapolations would make it easier to anticipate irrational currency wars, unsound consumer behaviour and an unconscionable race for resources (with more being done to craft a global financial system which feeds-back cybernetically in the interests of *Homo aequipondium*).

- *Homo aequipondium* the tool-maker

Construct a computer model using just a handful of economic variables, then keep adding variables (such as 24 hours trading, floating exchange rates, export subsidies, and so on). Before long, one arrives at a structure custom-built to bewilder populist politicians. The problem – if confounding populist politicians could ever be described as such – is that different parts of an economy operate at different speeds, such that a depreciating currency doesn't instantly generate increased foreign sales, any more than rising aggregate demand automatically conjures-up additional capital and labour. If one accepts that the global financial system ought to serve the political middle-ground, feeding-back negatively around political norms and socioeconomic set-points, then such a system *should* appear exasperating in the hands of *Homo sapiens*.

The question arises as to how a global financial system, which many see as neutral or amoral, could ever favour the political middle-ground? Such coevolution, by means of **unnatural fiscal selection**, is conceivable, because balanced-minds, supported by sound data, are able to sustain **economic equilibrium** by means of measured negative-feedback (see Fig. 15). In contrast, political pathologies, proceeding on the basis of **unsound induction**, consistently arrive at unstable economies, possessing toxic levels of positive-feedback. In effect, the proposed global financial system renders radical economies disordered, rather than having the 'ecological whole' disordered by those feckless radicals. Moreover, globalization is key, as regards crafting and sustaining that dynamic monetary equilibrium.

Conceivably, an **automatic stabilizer** would reduce the Centre-ground's sensitivity to supply and demand perturbations, whilst an **automatic destabilizer** would increase Left-wing and Right-wing sensitivity to the same. Moreover, NEURON would amplify that sensitivity in subordinate economies which are guilty of aggravating climate change, exhausting natural resources and accelerating habitat loss. Consequently, in addition to **stratified**

equity, whereby factual economic data becomes more readily available as one ascends through biased, logical, sound and balanced, one also encounters **lateralized equity**, whereby deviations away from the political Centre meet with increasing economic vulnerability (see Fig. 16). Wherever possible, such feedback being digitally-automated, without recourse to hopelessly ineffectual treaty obligations.

Figure 15: **Global financial system**

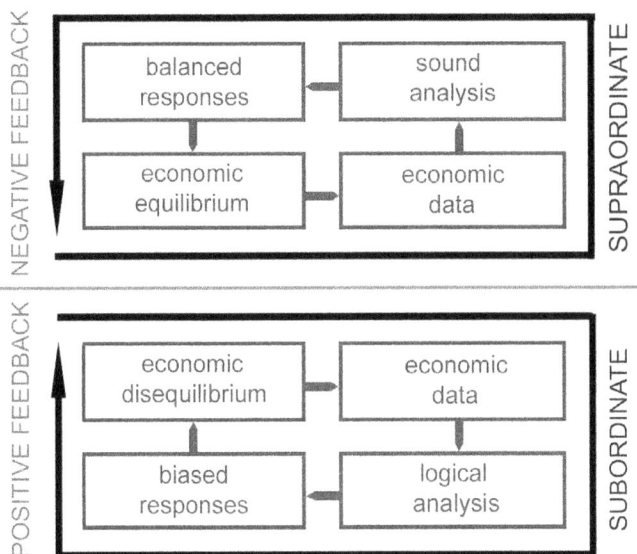

Unnatural fiscal selection is conceivable due to the impact of positive and negative feedback - feedback which spans balanced responses to sound analysis and biased responses to logical analysis.

Revised and updated, the global financial system would make ideology, politics and the military seem like a conceit, in-so-far as those clinging to those outmoded sources of power might assume they have choices. In reality, what choices they have would be more than matched by the selective advantaging of aequipine, at the expense of sapient (this being the least which must be done, if telepathically-induced effects are to be distanced from devotees of valid and invalid proposals). That socioeconomic advance rewarding *Homo aequipondium* with **absolute prosperity** and *Homo sapiens* with **relative prosperity** (relative prosperity denoting an accumulation of capital within an unsustainably subordinate position and absolute prosperity denoting a renunciation of capital within an enduringly supraordinate alternative).

Figure 16: Stratified and lateralized equity

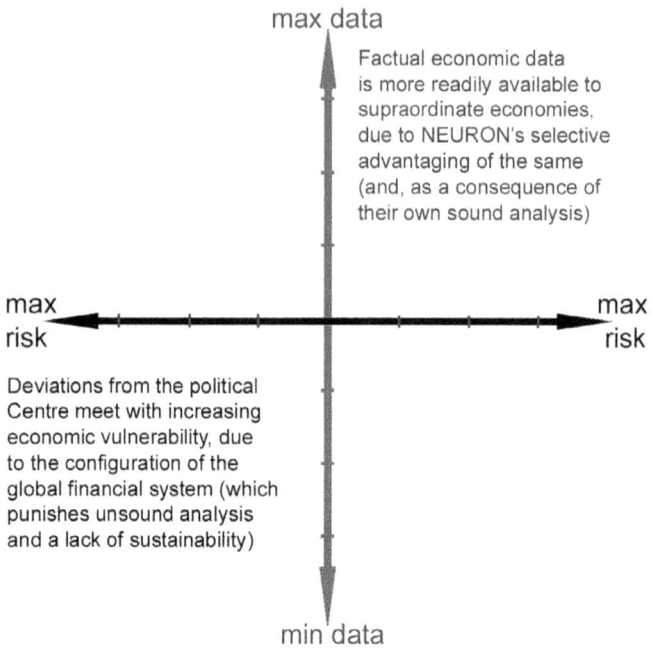

max data

Factual economic data is more readily available to supraordinate economies, due to NEURON's selective advantaging of the same (and, as a consequence of their own sound analysis)

max risk ← → max risk

Deviations from the political Centre meet with increasing economic vulnerability, due to the configuration of the global financial system (which punishes unsound analysis and a lack of sustainability)

min data

- Relative and absolute prosperity

Revised and updated, the global financial system would force an executive branch of government to question its doctrinaire insistence on **parsimony** or **profligacy**. Only capitalism tempered by socialism would be able to fabricate absolute prosperity, and with it conditions transcending pure economic well-being – unregenerate capitalism, like undiluted socialism, delivering only relative prosperity. This means that *Homo aequipondium*, due to its supraordinate status, would be well-placed to play-off subordinate antagonists, be they capitalist or socialist. In fact, all who are subordinated may find themselves serving the cause of cruciform advance, and with it a prosperity far-exceeding mere affluence. After all, what is the equilibrium price of truth, justice, liberty and clean air?

Some might accept subordination, in return for a short-lived relative prosperity, whilst others pursue an enduring supraordinate standing, even though it leaves them relatively poor. Absolute poverty isn't remediable through market forces, only through socialism's decomposition of unethical, unmerited and unproductive capital. As for absolute prosperity, that can only be achieved through the decomposition of power. Accordingly, those wielding a neuro-cognitive capability would, with the assistance of mainstream allies, decompose power and capital in order to alleviate absolute poverty and kindle absolute prosperity (increasing absolute prosperity and decreasing absolute poverty being the political middle-ground's professed goal).

The unsound, but otherwise humane, pursuit of wealth is tolerable – a case of *Homo sapiens* creating wealth for reasons beyond its narrow mercenary comprehension, and before cruciform coevolution renders the wider economy unfathomable to the same. Economic integration, particularly within federal unions, creates considerable economies of scale, enabling supply to comfortably outstrip demand, whereupon prices fall precipitously. Only *Homo aequipondium* has the ability to capitalize upon that kind of economic success – such success being wasted on *Homo sapiens*, which foolishly restricts domestic supply, whilst flooding the global market, forcing protectionist interventions upon friend and foe alike. That is to say, unlike **sapient economics**, **aequipine economics** actually warrants speculation.

Ch9: **Freedom of expression**

•   Psychodynamical succession

The influx of over 250,000 Palestinian refugees into Lebanon (following Israel's creation in 1948) proves that unchecked immigration can be destabilizing. In spite of initial economic optimism, sectarian differences soon arose between Christians, Sunnis, **Shi'ites** and Druze, due to the sheer number of Palestinian refugees living in squalid camps. Tensions escalated further when the Palestine Liberation Organization (PLO), led by Yasser Arafat (1929-2004), established paramilitary bases within Lebanon's Sabra and Chatila refugee camps, which they ruthlessly policed with their own militia. Then, in 1975, Lebanon erupted into civil war, with the PLO drawn into the conflict on the Muslim side. What followed, were repeated sectarian massacres – be it Palestinians by Christians, Christians by Druze or Jews by Palestinians (the PLO undertaking cross-border terror attacks on Israel).

The First Israeli Invasion of Lebanon, in 1978, amounted to a full-blown military operation aimed at countering this PLO threat. Additionally, Israel sought to support its Christian allies – and, if possible, quell Syrian influence in the country. It was against this backdrop that the USA initiated a peace process, aimed at reconciling Egypt and Israel – which resulted, in 1979, in the Egyptian-Israeli Peace Treaty. This treaty, which bore Egyptian President Anwar el-Sadat's (1918–81) signature, was perceived by the **Arab League** (who'd initiated Syrian involvement in Lebanon, some say with Soviet support), the PLO (who were dedicated to Israel's destruction) and Muslim factions (within Egypt's own armed forces) as a betrayal of displaced Palestinians. Thus, Sadat's summary assassination, by unsound factions within his own military, became symbolic of the Middle East entering a new deadly phase – one driven less by standing armies, and more by guerilla tactics, kidnappings and intifada.

More PLO attacks on Israel followed, leading, in 1982, to a Second Israeli Invasion of Lebanon. This time, however, the Israelis pursued outright regime change, the removal of Syrian forces, the installation of a Christian government and the sponsoring of pro-Israeli militias, with a view to smashing Palestinian terrorist cells. In the event, the assassination, by bomb, of the Christian Phalangist leader Bashir Gemayel (1947–82) ignited a wave of sectarian loathing – leading, in turn, to the massacre of hundreds of Palestinian males in the Sabra and Chatila camps, where logic suggested the PLO were most concentrated. This **Arab-Israeli conflict** owes much to the mind cycle, whereby each killing feeds-back positively, causing the whole to edge towards insanity. In fact, so crude was the bias and logic on all sides at that time, that extrapolating would've been easy and simple.

Palestinian Intifada, within the occupied West Bank, Gaza Strip and East Jerusalem – together with Arafat's proclamation of the State of Palestine, in 1988 – hint at the conflict becoming one of diplomacy, insurgency and counter-insurgency. Ominously, since the Second Israeli Invasion of Lebanon, a new peril has arisen – namely, Hezbollah (a fiercely anti-Western Shi'ite paramilitary organization, heavily financed by Iran). And so, is the Arab world merely biding its time, prior to liberating the occupied territories, as per **UN Security Council Resolution 242**? Or, are these massacres the new normal, as with Israel's killing of Gazans in 2018? Today, Israel continues to defy resolution 242, whilst provocatively building within the occupied territories – its martialing of logic betraying a weakness for cold-blooded responses, whose very predictability appears to be part of their strategic allure.

The question arises as to whether these unsound reprisals are enough of a deterrence to offset the cost of subordination? Of course, Israel's visceral responses represent a ruthless **retaliatory logic**, provoked by a persistent anti-Semitic historical bias. In an ideal world, one would replace *Homo sapiens* with *Homo aequipondium*, and start over – this time with boundaries rendered abstract by economic integration. That is to say, by allowing capitalism and socialism to arbitrate, the region could take on the appearance of a loosely-articulated federation – a federal union bound by raised living standards, social equality and disdain for ascription. More probable, however, is that the Middle East's sapient factions end-up being played-off against one another – until such a time as *Homo sapiens* becomes an archeological find

- Biblically illustrating secular and sacred

Cruciform sociometry, by collating coevolutionary data in six key areas, facilitates **cruciform calculus** (see Fig. 17). Imagine, for example, that we have four groups living in Jerusalem, labelled A, B, C and D (where group A is balanced-minded and politically neutral, group B is Right-of-Centre and sound, group C is unsound and Left-wing, and group D has a far-Right ideology). A member of group A might find themselves tried and condemned by group

D. Or, group A might enter into a coalition with group C, resulting in group B joining forces with group D. Far from proving that '*god*' and the '*devil*' are present in Jerusalem's day-to-day affairs, what this model actually illustrates is the pivotal role played by **biologically-produced** and **socially-constructed** factors (notwithstanding that '*god*' could be a social construct and the '*devil*' something irrefutably biological, making the interplay of the 'x' and 'y' axes anything but delusional).

Figure 17: Cruciform sociometry

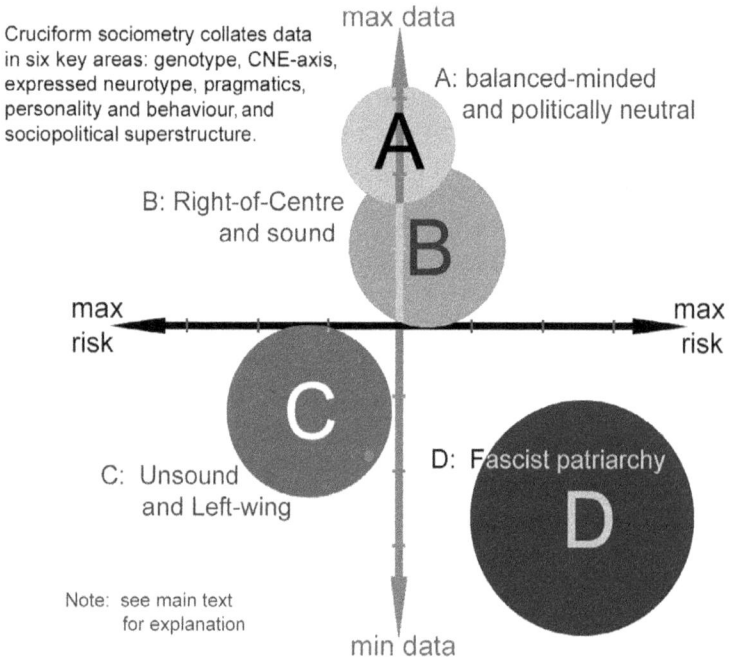

Cruciform sociometry collates data in six key areas: genotype, CNE-axis, expressed neurotype, pragmatics, personality and behaviour, and sociopolitical superstructure.

max data

A: balanced-minded and politically neutral

A

B: Right-of-Centre and sound

B

max risk

max risk

C

D: Fascist patriarchy

D

C: Unsound and Left-wing

Note: see main text for explanation

min data

Cruciform calculus possesses its very own **'uncertainty principle'**, in-so-far as it's possible to establish **psychostationary standing** or **psychodynamical momentum**, but not ascertain both simultaneously. That psychodynamical momentum – amounting to the way in which a given 'blip' is seen to move across the model – informing sociological predictions. For example, if group D (see Fig. 18) migrates down towards the right, expanding as it goes, it mirrors the rise of National Socialism in 1930s Germany. Clearly, knowing whether a group or individual is logical is one thing – ascertaining whether they're becoming more or less so, and how quickly, is quite another. As for the proposed global governance complex, that exemplifies **psychodynamical climax** – being comparable to a climax community, subject to stabilizing selection – save-and except that it utilizes stabilizing fecundity to maintain aequiponimity (the 'x' axis having created a 'y' axis in its own image).

Figure 18: Psychodynamical momentum

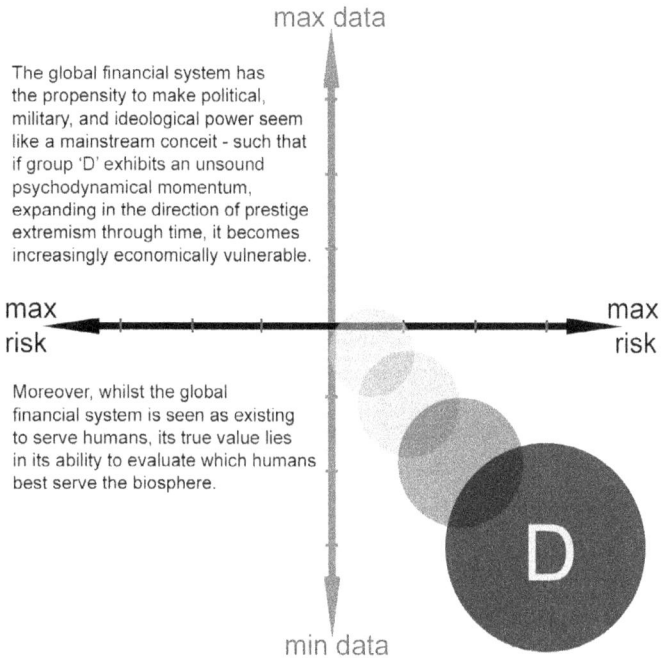

max data

The global financial system has the propensity to make political, military, and ideological power seem like a mainstream conceit - such that if group 'D' exhibits an unsound psychodynamical momentum, expanding in the direction of prestige extremism through time, it becomes increasingly economically vulnerable.

max risk ◄──────────────────────► max risk

Moreover, whilst the global financial system is seen as existing to serve humans, its true value lies in its ability to evaluate which humans best serve the biosphere.

D

min data

Balanced-minded groups and individuals have a **primary responsibility** to promote personal freedom, a **secondary responsibility** to decompose extremist power, and a **tertiary responsibility** to decompose capital (being aided in that aspiration by an equitable global financial system). A rich person's unsound concentration of power and capital crosses certain 'red-lines', sufficient to invite their subordination. Conversely, a person bereft of power and capital, having endured punishing maltreatment at the hands of others, might warrant supraordination by virtue of their balanced disposition – a scenario which '*biblically*' illustrates how a **supraordinate mind**, subject to unsavoury **subordinate behaviour**, can appear sublimely uncorrupted (a lesson, should the secular-minded need one, not to be drawn into basal responses at odds with textbook protocols).

- Antiquated Victorian broadsheets

Zionism is a logical consequence of biased anti-Semitism, and those who've exhibited that longstanding prejudice, be they German, Polish, Russian, Turkish or Arab, have provoked Israel into existence. Biased people, take note, you may one day meet with the logical ramifications of your own intolerance. A student of history, casually flicking through the broadsheets of the late 19th and early 20th centuries, will find report after report of anti-Semitic pogroms spontaneously arising in places such as Damascus (Syria, 1890), Odessa (Ukraine, 1905), Trakya (Turkey, 1934) and Kielce (Poland, 1946). The Third Reich (1918–45), which sought to profit from this deplorable sapiency, didn't devise anti-Jewish scapegoating, it simply harnessed an all-too-prevalent barbarism.

Indeed, one charitable explanation for the photographs in which spiritual mystic Grigory Yefimovich Rasputin (1872-1916) appears stood or seated with the last Tsar of Russia's family, is that Nicholas and Alexandra held in contempt those lowly subjects given to persecuting unorthodox religious minorities. In fact, those historians who promote the gossip, slander and hearsay surrounding Rasputin may be doing the deposed Russian royal family a grave disservice, as well as drawing attention to their unsound processing of historical information. Conceivably, those photographs show that the Romanov's were more sound than their proletarian subjects – and far more conscious than we of the terrors being meted-out on their soil. Terrors which would later be visited on the monarchy, aristocracy and bourgeoisie themselves.

Until recently, many of the news reports reaching The West from Sudan's Darfur Province read like sound-bites from those antiquated Victorian broadsheets – what with the now expelled President, Omar al-Bashir (1944–), ethnically-cleansing dissident Africans from Sudanese soil using murder, rape, looting, maiming and the destruction of property. Pro-Palestinians take note, both the Sabra and Chatila camp massacres appear tame, when compared to the sheer numbers who've been killed or displaced in Darfur by the Janjaweed *Arab militia*. Of course, the dilemma facing anyone participating in pogroms, ethnic cleansing and genocide is that subordination leaves them vulnerable to other people's retaliatory logic. That is to say, the unsound risk becoming '*hellishly*' embroiled in the biased and logical ramifications of their own diabolical thinking.

90

- The responsibility to protect

One could argue that Sir John Chilcot (1939–), former chairman of the Iraq Inquiry, seriously misled the British public with his 2.6 million word 'Chilcot Report'. After all, it's entirely feasible that his report gave confidence to oppressors, contrary to the spirit of the UN Charter. For example, one imagines that the recently ousted régime, responsible for the genocide in Sudan's Darfur Province, would've been nervously anticipating typhoon jets screaming across its airspace under the 'Gladstonian gaze' of Tony Blair (1953–), whose mindset closely mirrored the 'responsibility to protect' commitment of the **United Nations**. Begging the question, was the Iraq invasion (2003) initiated by balanced-minded parliamentarians, acting on sound information? Or was it, as some have suggested, an unlawful act of premeditated aggression, instigated by biased politicians?[21]

To be frank, if one presented a false case for ending the Holocaust, would that make ending the Holocaust a crime? At worst, it might amount to pardonable misconduct, while in public office. As an experiment, one could make an ill-informed case for evolution, just to see if evolution itself is denounced, by those legislators with an aversion to being misled. For the record, a sound conclusion, arrived at by unsound means, stands. Therefore, Tony Blair's crime was in adhering to the 'responsibility to protect' principle more than the United Nations. After all, if the UN fails to protect persecuted minorities, are those who volunteer to act criminals? In the future, there may be occasions when a person possessing telepathic abilities proves more worthy than a heavily-divided United Nations – much depends upon the complexion of the United Nations itself.

As regards critical thinking skills, when the unsound assert that the Iraq invasion occurred "*before all peaceful options had been exhausted*", it sounds to balanced minds like "*continue killing children, aged five years and under, with harsh embargoes*". In the event, the British Government brought about the removal of Saddam Hussein's (1937-2006) Ba'ath Party by means of elementary logic. Of course, that wouldn't permit those responsible to be fast-tracked into the heart of NEURON, but it would certainly make them invaluable mainstream allies, were that objective to be judged morally and politically correct. More broadly, if the UN obstructs actions aimed at preventing violent repression, it effectively mirrors

Left-wing incitement to commit such crimes – especially if those acting unilaterally find themselves maligned by asinine socialists.

In truth, anyone logical enough to see infanticide as the *'peaceful option'*, might just as easily contend that when Israel seized the Sinai Peninsula, back in 1967, they should've gifted it to the Palestinians, rather than handing it back to the Egyptians. Or, that displaced Palestinians should be given a *'193-state solution'*, amounting to them choosing any nationality, other than Palestinian. In the cold light of day, such logic appears outrageous, but the undue concentration of power and capital permits even preposterous forms of reasoning to be acted upon, hence its much-needed decomposition. However, it's permissible for logic and bias to be harnessed responsibly, especially where a balanced outcome is contingent upon preposterous happenings! That well-meant manipulation of mainstream subordinates being termed **cruciform coercion**.[22]

- *'A priori' coniunctum 'a posteriori'*

There are two types of unsound individual: (a) **rationalists** and (b) **empiricists** (rationalists inferring *a priori* from what they cogitate and empiricists inferring *a posteriori* from what they see). *Ergo*, unsound induction (see Fig. 19) comprises (a) or (b), whereas **sound deduction** comprises (a) + (b). Accordingly, hypothetico-deductive reasoning takes the rationalist's *a priori* working hypothesis and tests it empirically in an *a posteriori* manner. Cruciform calculus involves differentiating between unsound induction and sound deduction, so as to subordinate or supraordinate the groups and individuals involved. Basically, verifiable facts afford aequipine individuals an **absolute advantage** (born of them being more acquainted with an issue), whereas sapient individuals have, at best, only a **comparative advantage** (born of them being better placed to address an issue, albeit using unsound induction).

Accordingly, sound minds employ an *'a priori' coniunctum 'a posteriori'* approach, i.e. the (a) + (b) method. Inevitably, at the interface between sound and logical, one encounters **exceptionality**, **informality** and **colloquialism**. That is, beyond the prescriptive and proscriptive practices of applied cognitive pragmatics, one meets with general cognitive pragmatics and multi-dimensional deviance. Those seemingly sapient idiosyncrasies, in otherwise aequipine individuals – be they harmless assumptions or matters of taste – are attributable to the informality of the circumstance, rendering them psychometrically irrelevant. Naturally, categorizing an individual as '*less than sound*' or '*more than logical*' can be challenging, especially if they're only ever found within an informal setting. Notwithstanding that formal settings can also obscure neurotypic differences, due to sub-aequipine managerial incompetence.

Figure 19: Unsound induction (two principal types)

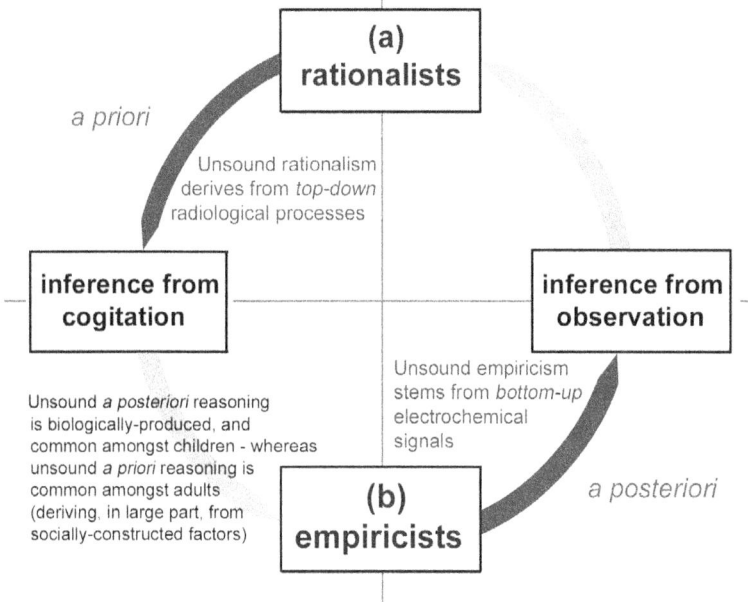

a priori

(a) rationalists

Unsound rationalism derives from *top-down* radiological processes

inference from cogitation

inference from observation

Unsound empiricism stems from *bottom-up* electrochemical signals

Unsound *a posteriori* reasoning is biologically-produced, and common amongst children - whereas unsound *a priori* reasoning is common amongst adults (deriving, in large part, from socially-constructed factors)

(b) empiricists

a posteriori

You'd be forgiven for thinking that puberty is some awkward transition from witless childhood to feckless adolescence – one in which the youth's burgeoning promise finds itself recklessly undermined by the unsound promotion of drugs, alcohol and self-image. Critical thinking skills, which might usefully counter those ill-conceived nostrums, habitually find themselves corrupted by **peer-to-peer communication**.[23] Evidently, it pays to positively-reinforce a critically-minded nature, if only to do justice to the teenager's 'axial potential'. After all, the principal reason why popular culture, local customs and religion contrast with science, is that the former avoid repudiation, much preferring proposals which feedback-positively, whereas science courts repudiation, placing negative-feedback at the heart of scientific advance (science working on the premise that if (a), which is falsifiable, cannot be disproved by (b), then (a) stands).

- Cruciform calculus

Typically, bosses burden their employees – seldom, if ever, streamlining their systems and procedures. That said, the proverbial *'fat and inefficient'* capitalist must work smarter and leaner, when faced with socialist demands for reduced working hours, paid leave, more pay, etc. Therefore, if Brexit Britain is to avoid debilitating itself with endless *'windbag additions'*, a strong socialist voice is called for. After all, antagonism, involving opposing voices, is the antithetical yardstick by which the political Centre is judged. However, such antipathy can be beneficial or damaging, with *Homo aequipondium* and *Homo sapiens* engineering constructive and destructive antagonism respectively. At its healthiest, contrasting views render the whole more competitive – and, in the case of commerce, shrewder, wiser and more resourceful. Thus, cruciform interventions encourage constructive forms of social and political divergence.

The biosphere comprises countless systems – ones married together prehistorically by ecology, but historically, by psychology. However, whilst human psychology continues to supplant ecology, anthropoid psychology remains prey to evolutionary laws. In other words, *Hominini* jeopardizes whole ecosystems, whereas cruciform coevolution raises the rationale of humans, sufficient to benefit both the environment and the people inhabiting the same. The academic study of humankind's relationship to the biosphere is termed **anthropobionomics**, with cruciform calculus steering humanity towards a rapport which is competitively aequipine, rather than competingly sapient. Therefore, replacing stabilizing and progressive selection with stabilizing and progressive fecundity makes things easier, not just on ourselves – but, more significantly, on the planet.

Throughout prehistory, plasticity has enabled human behaviour to be shaped by circumstance – as opposed to having conditions historically ruined by impressionable humans. As luck would have it, an open-competition now exists amongst *Hominini*, sufficient to determine the most competitive, rather than the most competing. **Socialization**, at its most competitive, comprises capitalism infused with socialism – such socialization wisely reconciling civilization's foremost socially-constructed antagonists, whilst all-the-while reinforcing the notion of a political middle-ground. Assisting the same, is cruciform calculus, which cybernetically-supports the most axially-advantaged, thereby appeasing unseen forces (such astrophysical exogeny shaping the coevolutionary endogeny in ways which rebut the anthropocentric hypothesis that 'x' is scarcely more than a structure and 'y' little more than an agent).

PART IV

Open-Competition

Chapters 10-12

Ch10: **Conceptual states**

• States that matter

Solidity, **liquidity**, **gaseousness** and **vapidity** emerge intergalactically due to the synthesis of protons, neutrons and electrons in deep-space, where electrostatic attraction is strongest (making them **corporeal states**). Conversely, biased, logical, sound and balanced emerge galactically due to the interplay of electric and magnetic fields, whose strength weakens electrostatic attraction (making them **conceptual states**). Temperature, by virtue of comprising electric and magnetic fields, conceptually regulates those **states of matter** mentioned above (not least, by being inversely-proportional to electrostatic attraction). Thus, the human **brain** is 'conceptualized' by forces arising well beyond the individual, leaving the **mind** to then dominate its inner workings. Not only are the brain's particles subsidiary to its electric and magnetic fields, those fields drive forward events, leaving consciousness in their wake.

Clearly, there are forces beyond human agency which might loosely be described as '*spiritual*' – available energy being the obvious example. Consequently, those applying telepathically-induced effects are, themselves, subject to '*spiritual forces*' beyond their personal control – the self-same electromotive and nucleomotive forces which long-ago brought life into existence. It's remarkable, but the universe's expanding electromagnetic spectrum has acquired the means to forge life. So, is life itself evolving, or the forces which shape life? And, when asymmetries arise which are at odds with applied ethics, does that conceptually-differentiating reality spontaneously correct such failings? This book argues that life is a manifestation of the electromagnetic – and therefore a reflection of something which, even today, appears largely inscrutable.

Without doubt, much of that '*heavenly influence*' derives from the fourth law of thermodynamics, whose very immutability enables some '*invisible hand*' to manipulate the corporeal to its insentient '*will*'. All of which begs the question, is '*god*' a universe which favours balanced-mindedness; only sparingly punishing its absence, and only so far as the physical and chemical limits demand? And, what if science found weak evidence showing that aequiponimity is favoured by natural laws, would we then be moving cautiously towards confirming '*god's*' existence? Certainly nature abhors destabilizing asymmetries, bringing much of human behaviour into question. But isn't that the point, in-so-far as ecosystems edge towards stability, whether humans comprehend the same or not. That what looks like a '*supreme being*' is, in point of fact, evidence of *Hominini's* lowly ecological status.

Sociologically-speaking, there are strong reasons for believing that logical and biased thinking are punished, and that sound and balanced thinking are rewarded. For example, logic demands that socialists absolve lawbreaking victims of circumstance, with an equal and opposing logic decreeing that capitalists incriminate the same (those on the Right-wing responding with a **bruising "*prove-it*" attitude** when faced with accusations of systemic bias, and those on the Left-wing, fired-up by social inequalities, militantly resolving to bring-down the establishment). However, whilst the unspeakable asymmetries conjured-up by *Homo sapiens* do provoke natural laws, those laws are unquestionably '*slow to anger*'. In other words, the aforementioned ideological antagonists will mutilate the corporeal beyond all recognition – prior, of course, to its sound and balanced restitution.

- Neurotypic differences

Deep-space nucleosynthesis is like resetting the universe's 'factory settings', because all those chemically-enriched galaxies, with their escalating spectra, suddenly find their emitted energy reconstituted in deep-space as the most basic quantum pieces. Those pieces later 'come alive' under the influence of *strong* and *weak* magnetic (**M.f/m.f**), electric (**E.f/e.f**), gravitational (**G.f/g.f**) and dispersional fields (**D.f/d.f**). Between them, those fields dictate the movements of each and every electron (e^-), proton (P^+), neutron ($N^{+/-}$) and photon ($ph^{+/-}$). From the perspective of the dark sciences, chemistry is concerned with available energy and physics with the laws governing its corporeal effects (see Fig. 20). Chief amongst such edicts are **Maxwell's equations** and **Coulomb's law** (although this work cautions against viewing distance as an independent variable, and maintains that electrostatic attraction is inversely-proportional to energy).

Contemporary taxonomy, otherwise known as the Linnaean system, classifies all known life hierarchically, i.e. domain, kingdom, phylum, class, order, family, genus and species. **Binomial nomenclature** exploits this acme of hyponymic hierarchies, in order to differentiate between living organism using only their two-part Latinate names, comprising the **genus** and **species**.

Whilst binomial nomenclature relies upon differences in phenotype, one is not bound by the same. Indeed, given that conceptual reality constitutes the universe's top-level causal influence, it would be perverse to overlook electromagnetism in one's classification of living organisms. Accordingly, this book introduces the term **neurotype**, as a means of distinguishing between species in terms of their cognitive processing of information. A cognitive ability which could have been predicted on the basis of those most basic quantum pieces.

Figure 20: Chemistry and physics

CONCEPTUAL
[available energy]

CORPOREAL
[electronic configuration]

Chemistry is concerned with available energy, and physics with the corporeal fabric which is subject to the same.

Thus, the kingdom Animalia, to which both *Homo sapiens* and *Homo aequipondium* belong, contains untold phenotypes, together with several key hominin neurotypes. Of course, this isn't simply a question of differentiating between human species, it's a question of understanding the ecological weight and importance of their respective axial responses (including reproductive strategies, conditioned behaviours, manufactured implements and ways of thinking). *Homo sapiens*, for example, conceived of the state, and looks set to utilize the same *ad nauseum*, even when more innovative social constructs present themselves. After all, with psychology currently supplanting ecology, shouldn't we be overhauling the Linnaean system, in order to anticipate the shocking environmental ramifications of radiating **neurotypic** differences.

• Inflexible autocratic states

When Mustafa Kemal (1881-1938) became the Republic of Turkey's first president, following the First World War (1914–18), his insistence on making Turkey secular was sound-minded. Secular, in the sense that Turkey abandoning **theocracy** and Arabic script, in favour of the Gregorian calendar, Latin typeface and a constitutional structure free of the prescriptive authority of Islam. A move which later enabled Turkey, under the Truman Doctrine, to gain both US financial support and membership of NATO (1952). This, together with multi-party politics and an unambiguous commitment to sexual equality, helped to forge a progressive Westernized state. However, this generally pro-Western state has since begun to falter, due to economic instability, recurrent refugee crises, Kurdish nationalism, Islamist revivalism, anti-secularism and repeated military interventions in its civilian affairs.

Consequently, Turkey has become an unstable democracy, whose collapse into autocracy may yet settle the argument as regards its application for EU membership. Today, under President Recep Tayyip Erdogan (1954–), the once banned leader of the Islamist AKP, one finds evidence of opposition purges, financial corruption, incarceration of journalists, human rights abuses, interference with the judiciary and an ongoing concentration of executive powers. Turkey (like Israel, and conceivably Great Britain) appears prone to defaulting into ruthless logic and unsound political practices, due to unalleviated external pressures and domestic disquiet. Proving, beyond question, that democracy isn't some inherently stable structure, it's merely representative – hence Turkey's worrying slide into authoritarian rule and Israel's draconian territorial stance.

In truth, representative democracy is promoted as being the optimal system of government, because a liberalization of society is thought to produce a more enlightened electorate. That is, a pluralist body politic, armed with a polymath-like grasp of domestic and foreign affairs, is thought able, through informed universal suffrage, to achieve social justice for all. More accurately, democracy is a telling psychometric device, eliciting information about a stressed electorate's rationale, with neocentric stratification offering the greatest conceivable stability within that otherwise mercurial arrangement. Such geopolitical entities arising due to the top-down influence of electromagnetic fields – an electromagnetic confluence which only rarely, if ever, stands the test of time.

- Fluid democratic states

The white supremacist National Party (NP), which governed South Africa between 1948 and 1994, erected an authoritarian superstructure termed apartheid. With all of its economic, political, ideological and military powers cynically concentrated in white hands, the black population became a subject people, victimized by habitual injustice. Needless to say, some black individuals fought back using the only weapon they had – namely, armed insurrection. As leader of the proscribed African National Congress (ANC), Nelson Mandela (1918-2013) possessed unparalleled distributional power – arguably greater than any activist before or since. His influence eventually galvanizing the end of apartheid, and with it the introduction of 'free and fair' elections. And so, in 1994, the ANC become the majority party within a government of national unity – one in which the National Party and Inkatha Freedom Party (IFP) both held cabinet posts.

With its executive, legislative and judicial functions wisely separated (both politically and geographically), a multilevel bureaucracy arose, founded upon national, provincial and municipal governance. Despite the pretence of multi-party politics, opposition to the ANC has been historically weak (with both the NP and IFP facing irreparable decline). Today, South Africa appears to be a largely top-down democracy, with constructive change being driven primarily by its ruling elite – its disaggregated political structure encouraging dialogue on issues as diverse as sexual equality, disabilities and youth. What's more, the economy has become largely inclusive, with an strong emphasis on self-employment, entrepreneurialism, job creation and access to education. Providentially, its pluralist fluidity might yet see the ANC split, over concerns such as recession, accession and secession.

On paper, at least, South Africa has transformed itself into a model democracy. And yet, in spite of striving to double domestic output, reduce inequalities and eliminate poverty, the country remains ridden with unsound behaviours. And not merely misplaced reasoning, but rather, disturbing activities made all the more alarming by the scarcely concealed attempts at normalizing the same. It would appear that *Homo aequipondium*'s efforts at promoting human rights and forging a world class legal system are being undermined by sub-aequipine elements. Take, for example, the Marikana Platinum Mine Massacre (2012), in which 34 protesters died. Many of the striking mineworkers, unhappy about pay and inequalities, foolishly began waving clubs and machetes. Needless to say, the South African Police Service, not noted for its balanced reactions, responded with live ammunition, killing dozens.

What the Marikana Platinum Mine Massacre shows, is that law enforcers aren't necessarily aequipine, and that repeated provocation may trigger sapient responses. However, whilst deep-rooted fracture lines open-up between rival human neurotypes, one must bear in mind that these fissures tear right through government, the police service and the races they serve. South Africa is an antithetical place, where even allies are torn-apart by conflicting rationales. Social action may have provoked a decomposition of white supremacist power, but it will take an overarching neuro-cognitive superstructure to fully capitalize upon that success. After all, "*that which is biologically-produced subverts all that is socially-constructed, unless that which is socially-constructed subordinates all that is biologically sapient*". Accordingly, pioneering lines are being drawn, with South Africa a telling template for the whole continent.

- **Volatile failed states**

Syria's recent descent into torture, decapitation and rape began as anti-government protests. President Bashar al-Assad's (1965–) attempts at quelling this insurgency ignited a countrywide insurrection, one which rapidly escalated along sectarian lines. It quickly transpired that the impetus driving this paramilitary uprising was the desire to police one's own territory, rather than advancing the insurgency in any clearly-defined direction. In practice, self-interested militias, like newly independent sovereign states, often find themselves manipulated by more powerful others. Thus, Syria faced the growing intrusion of opposing military blocs, one **Shia**, comprising Assad, Hezbollah and Iran (notionally allied to Russia), the other **Sunni**, comprising the Arab Gulf States and Turkey (notionally allied to The West). Of these, Turkey has since been rendered **schizoid**, by events on the ground.

The West's foremost problem has always been that the Russian-backed Shia are strong, as are the Arab-supported Sunnis, but that the so-called 'moderate rebels' are either too dispersed, strategically unreliable or closet extremists. Therefore, the conflict quickly became a face-off between factions possessing the tell-tale characteristics of *Homo sapiens*. One of *Homo sapiens* most telling features being its predilection for banks and ditches, amounting to so much entrenched tribalism. It was amid Syria's tribalistic retaliatory logic that the Sunni terror organization Islamic State briefly emerged, determined to unnerve its adversaries with coldblooded bias. Even with Islamic State now gone, bias and logic remain – such that no one knows for certain what a United Nation's **Security Council** ceasefire, transitional period and democratic elections would look like?

For the time being, it looks as though Hezbollah, Erdogan and Assad intend to shore-up the banks and ditches delineating their respective boundaries. However, biologically-produced factors will eventually overwhelm their given jurisdictions, as no social construct can withstand the cumulative impact of unsound behaviours – not least of which, their own. Ultimately, every group aspires to subordinate the competition, but that doesn't make them supraordinate. Supraordination, in the broadest sense of the term, is a judgement made on us all by forces transcending human agency – forces which 'conceptualize' beyond our narrow comprehension. Therefore, strengthening sovereign boundaries is all very well, but only by softening them with sound and balanced collaboration can progress be assured – that primitive love of walls revealing a weakness for authoritarianism.

• Gause's principle

Gause's principle states that no two species can occupy the same biological niche, a principle which extends to radiating forms of humanity – the winner being the most competitive, rather than the most competing. Humanists fail to grasp that **competitiveness** isn't decided by humans, it's a conceptual judgement arising beyond ourselves (with cynical interference, systemic bias and social learning merely delaying that judgement). Cruciform calculus harmonizes with nature, in-so-far as its promotion of sound and balanced thinking stabilizes communities. Conversely, bias and logic generate unsustainable asymmetries, which nature itself actively discriminates against. Given the unalienable nature of Gause's principle, an **epistemic community** of balanced-minded individuals, heavily-supported by factual data and possessing radiofrequency technologies, may be humanity's only hope.

Russian biologist Georgyi Frantsevich Gause (1910–86) experimentally demonstrated the principle of **competitive exclusion** using two closely-related species of unicellular paramecium. When grown independently, in a restricted space with strictly limited resources, both exhibited identical 's' shaped growth curves. However, when grown together, one was entirely eliminated. Much the same principle applies to radiating forms of humanity, when faced with diminishing space and declining resources. So, as *Homo aequipondium* and *Homo sapiens* vie for our restrictive planetary niche, the latter will be eliminated. Consequently, we needn't pray for balanced-mindedness to flourish, it could scarcely do otherwise (notwithstanding, that it would be desirable to have differential fecundity determine the outcome, rather than differential mortality).

Advanced societies have benefited enormously from an epidemiological transition, in which mortality shifts from being a consequence of infectious disease, compounded by warring, to being one of mostly degenerative illness. In the short-term, *Homo sapiens* risks turning that transition on its head, with fatal consequences. Indeed, some form of legal **extraterritoriality** becomes essential, once one accepts that exogenous factors afford sub-aequipine thinking an incurably finite evolutionary lifespan. That extraterritoriality effectively distancing *Homo sapiens* from the engine of global governance, and replacing its influence with verifiable facts and much needed negative-feedback. Here we enter a debate regarding normative controls, in-so-far as what ought to happen isn't necessarily debatable, and parliamentarians and their constituents must submit unconditionally using such plasticity as they can muster.

British sociologist Anthony Giddens (1938–) argued that social structures (otherwise known as structures or sociopolitical superstructures) are maintained by means of socialization, but that mainstream agents are free to ignore, replace or amend such structures, including the very nature of socialization itself. In other words, society has a **duality of structure**. **Structuration**, whereby the aforementioned social structures are revised or amended by mainstream agents, represents the global governance complex's Achilles heel. However, NEURON doesn't exist to be shaped by subordinate populations, engaged in pernicious social learning, it exists to determine how those subordinate populations ought to be shaped. So, whilst socialization instils the values of the political middle-ground, structuration betrays whether that message has been adequately comprehended and intelligently acted upon.

Ch11: **Progressive structures**

- Competitive structuration

As outlined previously, Anthony Giddens (1938–), disappointed that structural and social action perspectives were being presented as pure antagonists, presented his duality of structure thesis (whereby agents were said to influence structures, and *vice versa*, with neither having overall primacy). However, Gidden's duality of structure thesis has one principal weakness – namely, that the political perspective informs the sociological perspective, and *vice versa*, creating a spectrum of possibilities, be it inflexible autocracy, democratic fluidity or anarchic volatility (see Fig. 21). Therefore, structuration refers to the political middle-ground, and its absence to the extremes of anarchy and authoritarianism (only the Centre-ground possessing a comprehensive duality of structure). Consequently, any slide into despotism or disorder invites differential mortality, including an end to cruciform coevolution and that much vaunted epidemiological transition.

Figure 21: Normal curve of structuration

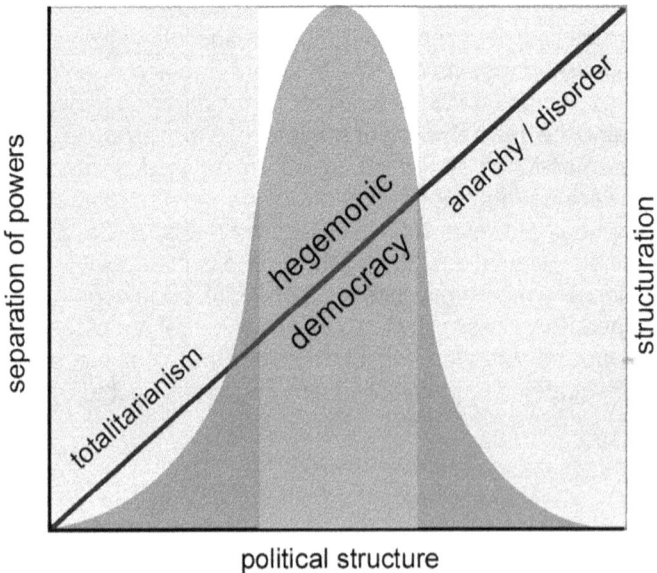

political structure

Within hegemonic democracy there's a duality of structure, as per Giddens (see main text). However, in the case of anarchy and autocracy, structuration diminishes

Using the 'structurating' vocabulary of Gidden's, mediation and contradiction have, respectively, a progressive and stabilizing effect on the disaggregated state, accelerating or reducing social change as a consequence. However, as sapient characteristics are being reproduced with every generation, many of today's social structures face an uncertain future, thereby prolonging the grief-ridden **anthropocene**. Accordingly, it's no longer sufficient to keep unsound elements away from telepathically-induced effects, they must be subordinated, in order to preserve the structures which guarantee their balanced application. Structuration is principally a mainstream phenomenon – one which permits an evaluation of society's multiplying agents. Soberingly, even when those **structurating** agents arrive at a progressive coevolutionary dynamic, the chances of it deviating sub-optimally remain high.

In retrospect, patriarchal structures have forced women into domestic roles – roles which have fostered balanced responses, primarily as a courtesy to their children. Or, to put it another way, *Homo sapiens* has unwittingly gifted children at least one well-balanced parent – women being more well-balanced, on average, than men. Meaning that balanced-mindedness predates coevolution, with no guarantee that the political middle-ground will be as successful at replicating that achievement. Much depends on the social structures themselves and how they impact upon the sexes. The danger, of course, is that misplaced structuration diminishes female aequiponimity, depriving women of their present **axial advantage** (or biological superiority). Fortunately, female subordination serves as a substitute for patriarchy, leading to a restitution of their balanced-mindedness.

- ## Socialization, acculturation and transculturation

In 2017, Muslim's were targeted in north London by a white assailant, in what became known as the Finsbury Park van attack. Without doubt, the attacker possessed a criminally-unsound mind, but was he driven by politically-inspired logic or simply blind prejudice? It just so happens that the now reformed Finsbury Park Mosque was the one were Abu Hamza al-Masri (1958–) proselytized as imam, prior to his conviction for inciting murder, and where Richard Reid (1973–), the so-called 'shoe bomber' (who planned to blow-up an American Airlines flight, in 2001), worshipped. It's conceivable, therefore, that the white assailant was motivated by unconscionable logic, rather than unintelligible bias – both **Islamophobia** and

Islamofascism feeding-back positively, with the aggressor's healthy mind paradoxically amplifying their hostility.

Bias notwithstanding, all of these attackers, bombers and orators were healthy, save-and except for the concepts of which they were cognizant. **Radicalization**, within a representative democracy, arises due to a failure of socialization, acculturation or transculturation. In the case of socialization, capitalist conditioning becomes infused with socialism, sufficient to reduce inequalities and advance the cause of emancipation. In the case of **acculturation**, an immigrant arrives in a representative democracy, whereupon they find themselves within a disaggregated structure, the impact of which makes them more open-minded and tolerant. In the case of **transculturation**, those representative structures are exported abroad, whereupon the overseas population becomes increasingly democratized, with their past culture more modestly expressed within the confines of their newly-liberalized structure.

As it happens, cruciform calculus doesn't subordinate individuals who've become radicalized – it subordinates biased and logical types, whose rationale makes them targets for radicalization. After all, biologically-unsound radicals remain biased and logical, in spite of **deradicalization**. Therefore, deradicalization is merely a palliative, whereas cruciform coevolution is a cure. That is to say, by pre-emptively categorizing biased and logical neurotypes, one can extrapolate as to who is likely to compromise due process, brutalize in custody, unlawfully detain or assimilate fanatically. Quite frankly, given the growing prevalence of telepathically-induced effects, it pays to anticipate parochialism, venality and corruption, if only to prevent their cruel and unusual application (with second-rate socialization, acculturation and transculturation compounding that concern).

- ## In God we trust

Take heart, unsound Americans will eventually go extinct. All the same, as a prospective cradle for *Homo aequipondium,* the USA possesses both strengths and weaknesses. Among its many weaknesses, one finds that a mere 4% of humans, who just happen to have been born American, feel compelled to consume a full quarter of the earth's natural resources. Moreover, to be elected president one must appeal to a large volatile electorate, often possessing subordinate levels of reasoning. Regrettably, one could all too easily arrive at a populist president, prone to actor-observer bias, who attributes other people's shortcomings to deficiencies in their character, but similar failings in themselves to *"rigged systems"*, *"widespread corruption"* and *"unjust conspiracies"*. What some might call a government of *Homo sapiens*, by *Homo sapiens*, for *Homo sapiens*.

Popularism, fanaticism and mass shootings all derive from deficiencies in socialization, transculturation and acculturation. But such is the pace of change, especially in The West, that the structures which the Western polyarchies export often make that political conversion ever more fraught. Accordingly, a person's psychodynamical momentum, when faced with unsettling change, may be in the direction of commendable aequiponimity or unintelligible sapiency. But, given that the world is populated by people who are manifestly *'more than logical'* or palpably *'less than sound'*, such a polarization is to be expected. Accordingly, the anthropocene is marked by the competitive exclusion of unsound Americans – many profoundly disorientated by other people's sound intentions and balanced ideals.

Presently, Russia and America appear compelled by an undemanding logic towards becoming opposing military blocs. Indeed, the world has already witnessed the rise of a **Russo-Shia bloc**, uniting the **Russian Federation**, Syria's Assad regime, Lebanon's Hezbollah and Iran's Islamic revolutionaries. However, Islamic State and al-Qaeda did a 'sterling' job of discouraging Western support for an **Americano-Sunni bloc**, centred upon the Arab Gulf States, by virtue of them decapitating and murdering Westerners. Who knows, perhaps it's Sunni insecurity which causes them to invest colossal amounts on armaments and munitions. Nevertheless, these two blocs remain putative adversaries in the **Iran-Saudi Arabia 'Cold War'**, with neither side holding a monopoly in respect of aequiponimity. Consequently, should it transpire that their aspiration is merely one of conspicuous consumption, *'pray god'* they mutually annihilate.[24]

Unquestionably, the Middle East is a finite space with strictly limited resources, making it the ideal setting for a *'heaven-sent'* solution. Mediation and contradiction, whilst much less reliable than *'god'*, do at least enable its citizens to steer their communities towards a competitive standing. What's important is that telepathically-induced effects are applied impartially, and that their usage transcends both domestic and foreign affairs. After all, what would it serve the Americano-Sunni or Russo-Shia blocs, if they allowed themselves to become crucibles for self-defeating sapiency? In other words, the proposed sociopolitical superstructure is motivated by the premise that humanity faces an uncertain future, unless non-negotiable **cruciform laws** are rigorously applied in a wholly independent top-down manner.

• Priceless antiquities

A Texas Department of Correction chief, exasperated at the killing spree of Bonnie and Clyde, once famously remarked "*put 'em on the spot…and shoot everyone in sight*". Islamic State provoked a similar logic with its abduction, torture and public execution of Syrian archaeologist Khaled al-Asaad (1932-2015). In addition to its destruction of priceless antiquities, Islamic State also committed, or threatened to commit, acts of extreme violence against Iranians, Iraqis, Israelis, Christians, Shi'ites and anyone questioning its disturbed credo. Islamic State's similarities to Nazism were telling – that is, capitalizing upon political instability, it used terror tactics and propaganda to advance its cause, whilst leaving mass graves in its wake. Proving, beyond doubt, that swaths of Sunni Islam would've been receptive to Hitler's anti-Semitic message – a fact not lost on the Fuhrer himself.

Later renamed 'Daesh', the decomposition of Islamic State's power is all but complete. Indeed, having lost the territory it once held several years ago – including the strategically important cities of Tikrit, Fallujah, Mosul and Raqqa – Islamic State now appears militarily defunct. This diminution of Islamic State is important, because – as many scholars have pointed out – Daesh's brutal expansionism risked bolstering the opposing **Russo-Shia bloc**, leaving The West with an ever more formidable adversary. Conceivably, Russia might have seen its assault on Daesh as an attack on Islamofascism, placing it on a par with its earlier mauling of the Wehrmacht. In fact, a simultaneous assault on an anti-Semitic fascist regime, by The West and Russia, has a strangely familiar ring to it – as does escalating tension between the victorious power blocs involved.

It's as though Daesh aggravated the longstanding Iran-Saudi Arabia 'Cold War', causing fissures to open-up between Russia and The West. Which, if nothing else, has brought into sharp focus the rationale of both Shi'ites and Sunnis, together with the respective merits of their Western and Russian allies. The temptation, of course, is to see this as a competition, replete with winners and losers. In truth, cruciform laws demand that all sides become competitive in their own right, sufficient to coexist amicably as part of a stable ecological whole – something which is unlikely to be achieved through reciprocal unsound behaviours. Capitulation *is* possible, provided one's rivals competitively reinvent themselves – with a global governance complex's support for *Homo aequipondium* facilitating constructive coevolution on all sides.

One way to provoke *Homo sapiens* is to introduce enlightened, possibly federalist, accords. That's because constructive détente inflames the sapient mind, leading to violent bloodshed. For example, there are sound-minded reasons for drafting accords in respect of Jerusalem, none of which involves actual peace, only the categorization of its denizens. Cruciform calculus, for its part, enables one to extrapolate as to the ramifications of various peace initiatives, based upon the prevalence of biased, logical, sound and balanced agents (with **aequipines** getting '*crucified*', due to the anthropocene's inescapably sapient dimension). However, abduction, torture and public execution notwithstanding, what unites both perpetrator and victim is the need to be competitive. Such that if Jerusalem's residents do arrive at a permanent political solution, it will be thanks to an invisible hand fostering a constructive, rather than destructive, antagonism.

• Russo-Shia bloc subordination

The Ukrainian Civil War (2014–) arose as a logical consequence of pro-European and pro-Russian forces pulling the country apart – pro-Russian separatists citing a weighted plebiscite as the legal basis for the Crimean Peninsula's summary annexation, together with purported cultural, political and linguistic ties. Such declarations of 'independence' aren't recognized by the Ukrainian government, or The West, who fear further land-grabbing on the part of the occupying Russians. Of course, it's conceivable that the EU is stronger for not having hostile elements bound within its jurisdiction, including outspoken Eurosceptic Brits. However, iron curtains rarely exist to separate aequipine from sapient, they exist to separate economic, political, military and ideological opponents.

Evidence of unsound thinking soon materialized, when, in 2014, Flight MH17 (*en-route* from Amsterdam to Kuala Lumpur) was blown out of the skies over eastern Ukraine. Piercing damage to the plane's fuselage being attributed by a team of international investigators to a Buk surface-to-air missile, brought-in from Russian territory and fired from an area controlled by Ukrainian separatists. With the cockpit wrenched from the aircraft, the plane continued along its fateful path, breaking apart as it fell (leaving all 298 occupants dead). Intercepted phone calls, traces of explosives, missile fragments, shrapnel holes and matching paint all served to incriminate the logically-minded Russians and their biased separatist cronies. By way of a response, the **US**, **EU** and **UK** all imposed sanctions against Russia (prohibiting long-term loans from being raised, together with travel bans, asset freezes and an embargo on the sale of arms).

Russia's diplomatic self-amputation was all but sealed when the United Nations, in conjunction with the Organization for the Prohibition of Chemical Weapons, concluded that chemical weapons had been used throughout the Syrian Civil War (2011–) by the Russian-backed government. Some might say that Syrian President Bashar al-Assad's actions hint at what his overthrow might lead to, given the stockpiles of chemical agents littering the country; whilst others might interpret these events as an admission that he's no longer in control of his own military. One such nerve agent, sarin, inhibits the action of *acetylcholinesterase* (an **enzyme** which deactivates nerve cells). Unsurprisingly, the victim experiences life-threatening positive-feedback, in the form of seizures, convulsions and respiratory problems (those cognitive, nervous and endocrinal spasms effectively mirroring Syria's very own positively-reinforced paroxysms).

In principle, both Washington and Moscow could be subordinated, due to unconscionable elements acting within their respective spheres of influence. Subordination being comparable to *acetylcholinesterase*, in-so-far as it serves a starting point for full-blown economic, political, ideological and military negative-feedback. Thus, anything incapacitating NEURON, or compromising its global neutrality, would have the hallmarks of a geopolitical nerve agent, convulsing the world in yet more spiralling sapiency. In other words, if such an eventuality were to occur, civilization would grow sick with accreted bias and accrued logic. The very antithesis, one might add, of **Russo-Shia bloc subordination** and **Americano-Sunni bloc subordination**, whereby their unsound cohorts meet with cybernetically-controlled decline.

What makes the EU so attractive, is that it's grasped that competitiveness is key. In contrast, what makes the **People's Republic of China** so worrisome, is that it appears to be gearing-up to selfishly compete. Making China appear pathologically ill-balanced and destabilizing of the international order.[25] This urgent need for principled self-interrogation arises at a time when the UK appears chagrined with many of the disaggregated debates previously cultivated within its hallowed universities – many inside the political establishment preferring unadorned practical proficiency, allied to a '*no questions asked*' partisan commerce. Of course, what makes this dialectical so pressing, is that every geopolitic comprises *Homo sapiens* and *Homo aequipondium*, and we need to be certain that our own political executive isn't deviating unethically.

- Biologically male or female

Females are well-balanced as a courtesy to children and males ill-balanced as an aid to externalizing territorially – save-and-except for the blurring effects of **autonomic** responses, steroidal hormones, social conditioning and remote interference. Nevertheless, strip away these subsidiary factors and one finds that women are, on average, more aequipine than men. It was the American psychoanalyst Dr Robert Stoller (1924–91) who argued that the expressions **male** and **female** pertain to **sex**, whereas the terms **masculine** and **feminine** apply specifically to **gender**. In the case of women, **aequiponimity** is typically a function of their sex and **sapiency** a reflection of their gender, with the reverse being true for men. The significance of which lies in the fact that the marriage of sound and balanced minds aids aequiponimity, whereas the marriage of biased and logical ones obscures the truth and disturb maturation.

As sex is biologically-produced and gender socially-constructed, it's possible to see how social conditioning might weaken women or educate men (or further enlighten females and visibly corrupt males). Thus, sex-obscuring **gender-fabricating variables** may combine to produce feminine males and masculine females (or **hyperfeminine** females and **hypermasculine** males). Given the possibilities, even experts in the field of **neo-Darwinian synthesis** would struggle to comprehend the coevolutionary significance of each and every gender-fabricating variable. What we can say, is that the human mind owes something to its initial starting conditions – because, although those conditions are little more than an inculcating positive-feedback mechanism, it's a mechanism which remains, almost without exception, anatomically male or female.

Thus, the human mind is a complex adaptive system, with gender an emergent property and sex an initial starting condition. As for radiofrequency technologies, they facilitate a non-invasive examination of foetal development, from the **zygote**'s most elementary absorption and emission, right the way through to fully-fledged adult consciousness. In between these two ontogenic extremes, the mind is subject to a continual stream of gender-fabricating variables. What makes women characteristically aequipine, is their ability to stay true to their maternal genome – for men, it's the ease with which any inherent sapiency is overwritten. Consequently, with patriarchies championing clear-cut territoriality and matriarchies favouring balanced nurturing, the actual difference between girls and boys is the difference between '*competitive*' and '*competing*'.

Ch12: **Battle of the sexes**

- **Sex-linked characteristics**

The feminist writer Ann Oakley (1944–) has argued that gender roles are socially-constructed, and that sex-linked characteristics can be overwritten by socialization. One could, by the same token, argue that social conditioning can be overwritten by sex-linked characteristics, with gender roles appearing biologically-produced. Looked at from the perspective of structuration theory, the Centre-ground's duality of structure leaves even well-intentioned socialization vulnerable to the impact of biologically-produced agents – agents whose axial responses radiate through time. Professor Oakley is half right, in-so-far as structures are the key to influencing mainstream agents. However, to avoid sub-aequipine elements subverting those well-intentioned structures, one must first distinguish between sound and unsound agents.

This book proposes that the biological origins of pecuniary and prestige extremism cannot be reasoned with – being, to all intent and purposes, a function of a given person's genotype. That while feminists perceive women as an oppressed group, their subjugation owes less to misogyny, *per se*, and more to inherited bias and logic. *Ergo*, subordinating sub-aequipine elements enables coevolution to be beneficially steered – so that, in spite of sex being scripted by one's genetic code and gender by semiotics, society's ability to script gender responsibly increases through time. Remember, cruciform calculus isn't used to subordinate misogynists, it's used to subordinate biased and logical types, whose reasoning makes them appear intolerably chauvinistic. Equally, **misandry** becomes less of a problem, by virtue of unsound women meeting with an equality of subordination.

Brain lateralization, for its part, is largely a function of sex (with women's brains appearing less lateralized, on average, than men's). Consequently, in the absence of socially-constructed gender roles, the brain's physical dissimilarities would inevitably affect the way in which men and women respond to stimuli, be it **hormonally**, **electrochemically** or **radiologically**. Pavlovian and operant conditioning – which, taken together, have afforded placental mammals a significant conceptual advantage over the last 65 million years – have since evolved in humans into cultural conditioning, gender identification and advanced learning. Clearly, a certain amount of plasticity is desirable, but is it wise to overwrite biologically-produced sexual differences? In truth, one may, or may not, wish to overwrite those differences, it very much depends upon the individual.

Ann Oakley's well-meaning attempts at expunging sapiency are, of course, doomed to failure. That's because, short-lived plasticity notwithstanding, the unsound have underlying cognitive, nervous and endocrinal weaknesses which cannot be reasoned with. Epigrammatically-speaking, "*agents subvert structures, unless structures subordinate unsound agents*". Consequently, competitive exclusion, in the case of *Hominini*, amounts to *Homo sapiens* compromising tried-and-tested sociopolitical structures, leaving *Homo aequipondium* to win through, but at great personal cost. Somewhat ineluctably, we're forced to conclude that subordinating society's sub-aequipine elements is paramount, if social constructs, like sexual equality, are to consolidate themselves. That is to say, feminists, such as Ann Oakley, aren't wrong – they've simply been appealing to the wrong species.

- Gender plasticity

The Centre-ground is home to an unstoppable **gender dialectic**, with the democratic state entering into a protracted self-interrogation regarding the true nature of male and female identity. Notwithstanding, that gender is afforded a low-priority in the minds of those facing the gravest of anxieties, with sex-linked characteristics 'kicking-in' in relation to territorial concerns and domestic nurturing. Nevertheless, the fact that gender, be it **masculine** or **feminine**, is socially-constructed, bears testament to the malleable – some would say, highly-indulgent – nature of human psychology (a malleability which comes in various guises, be it biased, logical, sound or balanced). Because an agent's sex-linked characteristics are determined biologically and their gender socially, there's a **duality of gender** – whereby **sex-linked** characteristics influence the gender dialectic, and the gender dialectic indulgently overwrites society's sex-linked characteristics.

Women are, on average, more aequipine than men, thanks to some long-standing social structures – some of which, radical feminists would've militantly abandoned long ago, e.g. the nuclear family. The challenge, if you're a man, isn't to mimic femininity, but rather, to possess the hormonal, electrochemical and radiological responses of a well-balanced women – responses which arose under ostensibly repressive conditions. As if taking inspiration from the same, cruciform laws don't consult men as regards a prospective reassignment of their sex-linked characteristics – they simply persist, until men arrive at a more enlightened psychostationary standing, ideally by means of coevolution. Of course, those cruciform laws would be subverted by the very males they're intended to 'advantage', were it not for the subordination of unsound, but otherwise socially unimpeded, mainstream agents.

Figure 22: Duality of gender

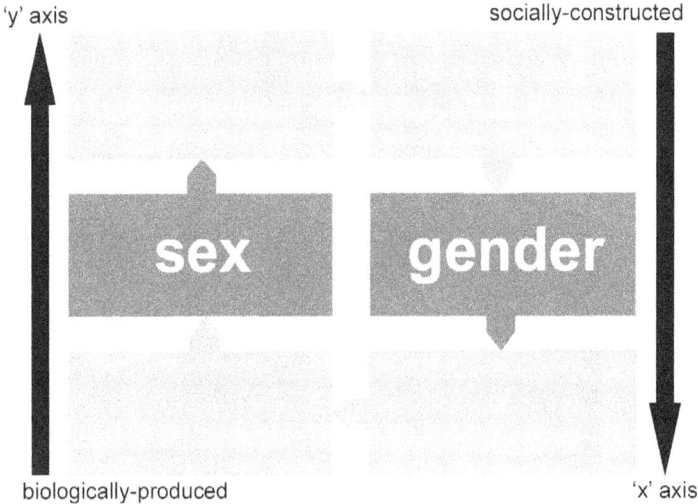

'y' axis   socially-constructed

sex   **gender**

biologically-produced   'x' axis

In order for sex-linked characteristics and gender to constructively coevolve, an appropriate gender dialectic is called-for within the balanced-minded political middle-ground

Cold logic demands unregulated mainstream structuration, sufficient to categorize those driving forward social change. Making cruciform sociometry a measure of structurating and **socializing** processes. Mathematically-speaking, cruciform calculus is obsessed with relationships – which, in the case of society's duality of gender, is the relationship between gender-identifying agents and gender-fabricating structures (see Fig. 22). The cruciform model appears outwardly inert, save-and-except for some arithmetical **function**, denoting a point's trajectory, when plotted onto that Cartesian **x-y coordinate system**. Constructively tweak the biological and sociological factors driving that point's migration, however, and society begins to perceptibly coevolve towards absolute prosperity – with sex-linked characteristics and gender both following suit.

The cruciform model's **vertical axis**, or **'y' axis**, denotes the agent, and the **horizontal axis**, or **'x' axis**, the structure (for

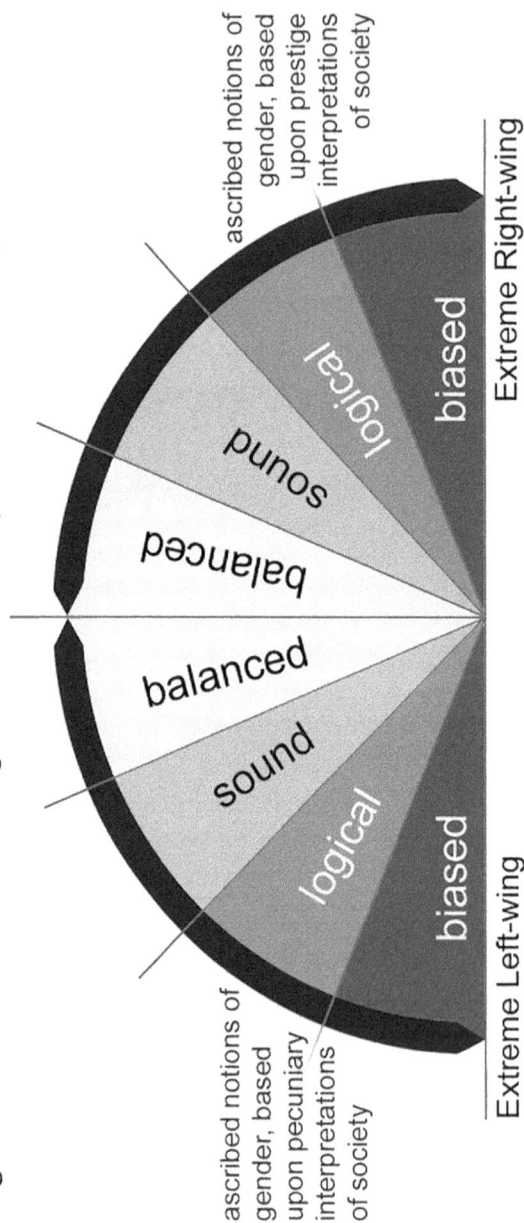

Figure 23: Gender convergence and the political middle-ground

An absence of adequate structuration, within ideologically-unsound societies, compromises the duality of gender commonly associated with the political middle-ground. Consequently, one encounters hypermasculinity and hyperfemininity within far-Right regimes (regimes which ascribe gender roles, amid widespread sexual inequality). The radical-Left, conversely, is far more complex, often promoting hypofemininity, amid hypermasculinity (thereby maintaining a comparative gender divergence).

simplicity, we've used the terms 'agent' and 'structure'; but within the context of this book, the terms 'neurotype' and 'superstructure' would be more pertinent). What makes the above relationship so arithmetically fluid, and hence mathematically fascinating, is that *superstructure* isn't an altogether independent variable, any more than *neurotype* is a purely **dependent variable** – the political middle-ground's duality of structure makes them **mutually dependent**. Significantly, as a given curve, function or line moves away from the vertical 'y' axis, that fluidity is replaced by an inflexible dependency or anarchic absence of association. Such that the sapiency driving that coevolutionary collapse is hastening its own competitive exclusion, by failing to maintain the mechanism necessary for its own survival.

- *Tabula rasa*?

The human mind's assimilation of declarative and procedural knowledge, from an initial neonatal 'blank slate' (or ***tabula rasa***), has been speculated upon by the likes of Aristotle (384-322 BC), Thomas Aquinas (1225–74), John Locke (1632-1704) and Jean Piaget (1896-1980). That 'blank slate', or mind cycle, certainly starts 'blank' – in-so-far as it lacks inculcation, experience and conditioning – but as the fabric of the brain dictates what can reasonably be inscribed, 'slate' is something of a misnomer. A '*uniquely-blank, person-specific material, imperfectly suited to inscription*' would be closer to the truth. Tellingly, those who stray beyond coevolution meet with natural selection, differentially evaluated on the basis of radiological, electrochemical and hormonal activity. Needless to say, that pathological deviation terminates the middle-ground's gender dialectic, and replaces it with something altogether more objectionable.

Here, we're faced with the thorny question of whether gender identification should be encouraged or discouraged (see Fig. 23). In truth, gender identification becomes more pronounced as one descends the vertical 'y' axis, through balanced, sound, logical and biased (biased societies prejudicially ascribing male and female characteristics; with **hypermasculinity** and **hyperfemininity** appearing 'self-ascribed' in those instances where biased minds meet with a never-ending stream of gender-fabricating variables). All of which makes a more muted masculinity and femininity appear sound. However, once we arrive at balanced-mindedness, the objective truth, as it pertains to each and every uniquely-inscribable man or woman, is so multifaceted – particularly in relation to extrinsic priorities – that the very notion of gender has a much reduced currency.

Gender plasticity, like hominin plasticity in general, remains controversial, due to the fact that much of what is assimilated is at odds with behavioural imperatives. That said, gender identification is a fundamental freedom, and essential to the dialectic which the Centre-ground hopes to encourage. Therefore, somewhere between outright 'sexual apartheid' and the 'genderless telepathic fusing of male and female consciousness', lies the hotly-contested boundary between sapient and aequipine (**gender convergence**, built upon sexual equality, denoting aequiponimity, and **gender divergence**, built upon sexual inequality, denoting sapiency).[26] Ultimately, the question is not whether one is male or female, or even whether one is masculine or feminine, but rather, whether one is sound or balanced – unfortunately, a lifelong preoccupation with sex and gender compromises both.

- Sapient divergence

Feminists rail against inequality, not realizing that sexual discrimination arises as a logical consequence of our evolutionary past. In other words, *Homo sapiens* has proven to be an extraordinary specimen, for whom society and family were always going to be patriarchal institutions, custom-built around asymmetrical foundations. Moreover, in the absence of aequiponimity, that chauvinistic hominin would've endured until the cumulative impact of unsound behaviours threatened its very existence. Today, however, *Homo sapiens* has more than its own hubris to contend with – it's faced with competitive speciation, born of adaptively radiating biological processes. Therefore, feminists need to stop railing against that transitive taxonomic specimen, and start soundly promoting its aequipine replacement.

Sociologists are conflicted, precisely because humanity is at a crossroads in human evolution. Accordingly, academics must now qualify their arguments, and specify whether their interpretations pertain to sapient or aequipine individuals. One such academic, who failed to clarify which stratum of society her published theories pertained to, was American psychologist Sandra Bem (1944-2014). Bem's **gender polarization** thesis maintains that socially-constructed notions of masculinity and femininity find themselves habitually exaggerated, whereupon gender roles become mutually exclusive, such that to be feminine a woman must reject all things masculine, and *vice versa*. Logically, this propels men into highly-paid public and political roles, and compels women to retire into unpaid family-orientated domestic servitude.

Sandra Bem was, of course, skillfully deconstructing *Homo sapiens*. *Homo sapiens'* subordinate mental processes tending towards gender divergence, gender stereotyping and sexual exploitation. By way of example, Palestinian males are highly-valued within the occupied West Bank, Gaza Strip and East Jerusalem. However, divergence, stereotyping and exploitation within those occupied territories leaves many Palestinian women feeling dangerously undervalued (what Bem, given her 'single species' portrayal, would've termed gender polarization). Consequently, UNICEF estimates that up to two-thirds of murders, occurring within those occupied territories, are acts of culturally-sanctioned homicide, i.e. 'honour killings'.[27] Bem saw gender polarization as driving sexual oppression, whereas this book contends that gender divergence is actually a conspicuous side-effect of inborn bias and logic.

Clearly, the question arises as to whether society should permit gender divergence to be 'exported' to The West, given its association with female infanticide, acid violence, genital mutilation, bride burning, dowry killing and death by stoning? Moreover, the fact that such violence crosses religious divides suggests that credo alone isn't responsible, but rather subordinate cognitive, nervous and endocrinal processes. Academically-speaking, our age **furnishes** and **furbishes** (equipping and contemporizing, by turns). Clearly, Professor Bem's theories need revising and updating, in-so-far as *Homo sapiens* will cause gender to diverge and politics to polarize. That said, cruciform laws exist to counter destructive forms of sexual and political antagonism, eventually bringing about a state of pioneering convergence – not least, the constructive merging of male and female consciousness.

• Divine retribution

Religion isn't the root of all '*evil*'. If it were, '*evil*' would've been absent prior to religion, and atheists would be above suspicion. What religion does do, however, is propagate logical arguments, incorporating irrefutably false premises. Therefore, religion is a common source of subordinate reasoning, but by no means the only source. Faulty reasoning, allied to ill-balanced reactions, is actually the root of all '*evil*' – which one finds evidence of in both atheists and theists alike. Indeed, it's delusional to have faith in *Homo sapiens* – and any humanist who does, hasn't grasped its shortcomings. You see, *Homo sapiens* is beyond socializing, sufficient to avoid '*evil*'. Consequently, even if one were to eliminate religion, idolatry and worship, there'd still be a million other concerns to contend with. To eliminate upwards of one million problems, extirpate sapient axes.

As luck would have it, *Homo sapiens* can be played-off against itself, its penchant for despotism and anarchy betraying an innate capacity for mutual distrust, political divergence and territorial conflict. Countering this genotypic dysfunctionality, while preserving the balanced female **psyche**, necessitates an immutable global superstructure. In other words, Middle-Eastern males may one day meet with non-negotiable cruciform coercion within the confines of their domestic political environments. The crunch comes, of course, when they realize that their culturally-sanctioned sexual tyranny has subordinated them, leaving them feeling strategically outsmarted and intellectually worthless. The choice is stark – coevolve as equals, or face evolution in the raw.

Clearly, we can all have faith in the **'one true' competitive exclusion**. Not least, because of the demands it makes on those fortunate enough to be classified as supraordinate. In practice, cruciform coevolution can be an unpalatable process, but one which compares favourably with the disastrous alternatives. Propitiously, having displaced unsound thinking, society will have cured itself of countless other concerns, making the pain all the more bearable. Think of it not as targeting gun-owners, *per se*, but rather the irrational basis of gun ownership – the conspicuous bearing of arms serving to draw attention to the owner's defective rationale. Of course, *Homo sapiens*, particularly one that's armed, is difficult to argue with – just how difficult, one can only extrapolate!

- **Women impart**

American **polemicist** Warren Farrell (1943–), in his book '*The Myth of Male Power*' (Simon & Schuster, 1993), asserted that neither sex has a monopoly on power – or, as he logically expressed it, full control over their lives – such that the sexes were said to appropriate societal roles, risking death in childbirth versus death in war. As an argument, it's barely distinguishable from the one promoted by Daesh, who saw women as having predetermined responsibilities within the home (ones which carried comparable risks to Islamist males severing heads on the IS front line). More apt, would've been a book entitled '*the myth of male aequiponimity*', because, whatever those designated duties happen to be, and regardless of where real power lies, males are consistently less sound. Moreover, the structures which concentrate power are typically patriarchal – quite unlike the matriarchal empowering of others.

Whether they 'appropriate roles' or are 'ascribed responsibilities', men and women are both plausible victims of ascriptive practices. However, if one defines power as the probability of a person or group carrying out another's will, even when opposed, as per Max Weber (1864-1920), then the male obsession with firearms suggests that the 'will to power' is that much greater in men (as is the determination to concentrate power, often for its own sake). Again, Farrell's argument falters, due to patriarchies habitually *"favouring punishing; using reward selfishly, and without obvious limits"*. Therefore, it comes as no surprise to find that women have been historically disadvantaged – economically, politically, militarily and ideologically – leaving them bereft of every one of these principal sources of power.

It's no coincidence that the best of women's instincts conform to a rule taken from childcare – namely, *"favour rewarding; use punishment sparingly, and only to define the limits"*. Of course, a neocentrically-stratified patriarchy could be just as well-balanced as a matriarchy of comparable size. However, atypical communes notwithstanding, a demographic chosen at random would lend itself to sub-aequipine males concentrating power, often through sheer weight of numbers and an inborn competitiveness. The foremost determinant of an individual's neurotypic standing, whatever the prevailing stratification, is their **cognitive-nervous-endocrine axis**, or **CNE-axis**. As women possess an axial advantage, gender convergence, weighted in the woman's favour, underpins heteronnubial dualism – with the balanced merger of the writer and co-author's axes proving radiologically, electrochemically and hormonally illuminating.[28]

PART V

Emotional Content

Chapters 13-15

Ch13: **Cogito, ergo sum**

• Light transduces, therefore I think

Abelard's dictum – namely, that "*by doubting we come to inquiry; through inquiring we come to perceive the truth*" – leads to sound reasoning. In practice, if one casts doubt upon everything which can be doubted, however improbable the deception, one is left with the fact of one's own cognitive-nervous-endocrine axis as the most unassailable of truths – doubting one's axis being different to doubting one has an axis. Such **Cartesian soundness**, immortalized by the phrase '*cogito, ergo sum*' (I think, therefore I am), leaves open the possibility of deception. Sound individuals being certain of their feelings, but not necessarily the events which invite strong emotions. Such distortions, including state-sponsored disinformation, leaving unsound persons, both inside and outside the law, in possession of misdirected anger, revulsion and hostility.

Cartesian soundness suggests that **evidence**, as well as allegations, ought to be falsifiable. For example, video evidence, which, if fabricated, *could* and *would* be discredited, but isn't, constitutes irrefutable proof. Whereas all other forms of video evidence could, in theory, be faked. Lamentably, the law is ridden with bias and logic. For example, favouritism towards the defence or prosecution, in the case of judicial proceedings, constitutes unsound deliberation – with the law's standard texts regularly finding themselves corrupted by subordinate elements. For its part, systemic bias arises whenever the police, courts or judiciary identify exclusively with the prosecution (those highly-remunerated sub-aequipines experiencing misplaced anger, revulsion and hostility, due to their received learning being undermined by subjective social learning).[29]

In 2008, Aisha Ibrahim Duhulow (c.1995-2008) was convicted of adultery by a Somali Court, resulting in her being buried up to her neck and stoned to death – begging serious questions, as regards which specimen of hominin Islam hopes to attract. Unlike *Homo sapiens*, which is instinctively drawn to the bias and logic inherent in Sharia proceedings, *Homo aequipondium* self-interrogates regarding the probity of the law, the reliability of plaintiffs and the impartiality of judges. Such self-interrogation is crucial, as the ill-balanced application of remotely-induced effects amplifies congenital failings on both sides of the law, leading to spiralling sociological dysfunction. Needless to say, neocentric stratification engenders a first-class legal system, whilst providing for the balanced application of telepathically-induced effects.

Many in The West, keen to cultivate a hardline appearance in respect of the Russo-Shia bloc, have welcomed the prospect of an iron curtain 'ringing down' across eastern Europe. However, the influence of the Russians – in terms of cyber-attacks, nerve agents and land-grabbing – pails into insignificance, when compared to the unsavoury impact of The West's own unsound citizenry, especially if you happen to be in long-term social care due to a learning disability.[30] Cruciform advance must be a 'war' fought on all fronts, by non-partisan forces transcending mainstream human agency – a 'war' which liberates those with little or no voice. After all, learning disability has its corollary in teaching disability – with sub-optimal socialization and substandard structuration fostering institutional abuse.

If, in spite of a spirited détente, the Americano-Sunni and Russo-Shia blocs do come to blows, it will have less to do with the 'unspeakable east', *per se*, and more to do with both sides' subordinate factions gaining the upper hand. Failing conflagration, however, sub-aequipine elements will continue to exert a worrying domestic influence, placing a given country's own citizenry at the mercy of that 'enemy within'. Currently, the United Kingdom risks rewarding those extremist views responsible for the late Jo Cox's (1974-2016) politically-motivated murder. Unsound Brexiteers being more conversant with their own emotions, than the facts themselves. That's not to imply that their brain's aren't healthy, it's just that their minds feedback-positively, whatever the facts – and in spite of copious negative-feedback at the autonomic, endocrinal and supraordinate levels.

•   Coevolutionary dynamics

Society's genetic substructure codes for an infinite array of CNE-axes, supporting four principle neurotypes – neurotypes which are identifiable on the basis of their social constructs, idiosyncratic structuration and associated behaviours. That democratic codependence, synergistically comprising genes and memes, is what makes the political middle-ground so vibrant. Cruciform sociometry (which collates coevolutionary data in six key areas) is ostensibly fractal, in-so-far as neurotype and superstructure can be of variable scale (see Fig. 24). In terms of cruciform calculus, the sum of all the relevant biological factors gives one the neurotype, as marked on the 'y' axis (with the superstructure, which is the sum of all the pertinent sociological factors, being marked on the 'x' axis). That is, *"neurotypes derive from CNE-axes, coded for by genes, and superstructures derive from personalities and behaviours, coded for by memes"*.

Figure 24: The coevolutionary dynamic (genes and memes)

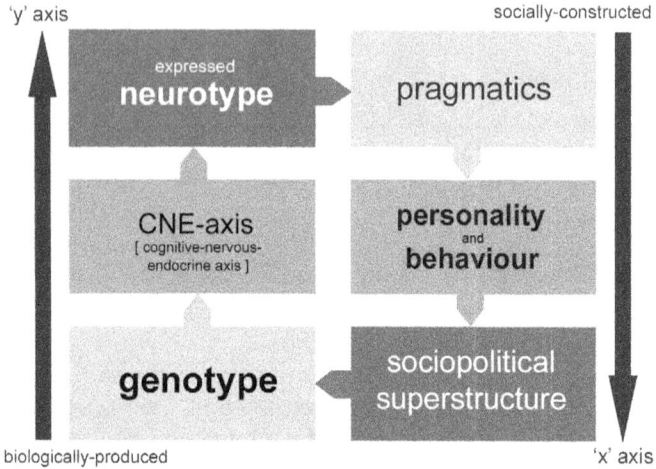

In the case of an individual, one is concerned with how their genotype expresses itself, sufficient to affect structuration (more broadly, one might examine how a given group expresses itself, and the structuration which then ensues).

The biological concept of the **nuclear family**, comprising a father and mother, plus their immature offspring (*no pun intended*), is a human universal. Beyond that nucleated family unit, however, symbolic language use has served to create the **socialized family** (which, like gender, is a social construct). In coevolutionary terms, not everything which is biologically-produced is good, and not everything which is socially-constructed is bad – there are, as many will attest, sound and unsound nuclear families, just as there are functional and dysfunctional socialized households. Whether nucleated or socialized, the family serves as a microsociological counterpoint to macrosociological constructs – replete with its very own scapegoating, incitement, patriarchy, insurgency and exploitation (making a decomposition of power and capital within the home, every bit as important as its broader separation, fragmentation and denial).

Elections and referenda are constitutionally structurating, enabling a family to unburden itself of socioeconomic stress through the expedient of universal suffrage. Of course, the reverse is also possible, whereby the outcome of a given plebiscite unwittingly

128

burdens a family with insuperable stress, due to the unsound nature of the result. A political elite, prone to actor-observer bias, will dismiss hard-pressed families as having flawed dispositions, making structuration seem like hard work. Right-wingers often failing to grasp that capitalism is fraught with difficulties, which privileged advocates of the markets ought to sympathize with, rather than merely denigrate. Even so, like the archetypal dysfunctional family, the **two-party system**'s antithetical exchanges enable a citizenry to somehow claw their way towards civilization – albeit in the manner of some antagonistic gene therapy.

- Functionally-nucleated families

American sociologist Talcott Parsons (1902–79) argued that the nuclear family grew in significance, thanks to its functional compatibility with highly-developed economies. Parsons further maintained that the principal functions of the family were: 1) the primary socialization of children, and 2) the stabilization of adult personalities. Thus, progeny became inculcated in the norms and values of their culture, via primary socialization within the home, prior to them meeting with secondary socialization beyond it. As for the parents, they were said to benefit from companionship, stress reduction and emotional security. Predictably, Parson's 'functions of the family' thesis was attacked as naïve and unrepresentative, with some complaining that the nuclear family doesn't possess a monopoly on any of these professed functions.

In spite of the attacks, Parson's thesis stands as a **normative device**, his detractors failing to appreciate that the family's functions are normative, and have been ever since language first evolved. The fact that families experience dysfunction doesn't detract from the family's anticipated role. After all, most couples carry the 'functions of the family' thesis to the proverbial alter, believing that any dependents will become suitably socialized within a secure family setting. And, how wrong many of them prove to be, as bias and logic play havoc with the anticipated script. Here we enter a prickly debate regarding agents and structures, and whether one should criticize the structures or the agents. Basically, one shouldn't allow the shortcomings of *Homo sapiens* to undermine our appreciation of any social structure, including the family – after all, no structure is immune to their insurmountable deficiencies.

Feminists are correct, patriarchy *has* permeated family life, affecting the manner in which primary socialization and emotional stability play-out within a given household. In fact, women haven't just become more sound and balanced as a courtesy to their children, they've also found themselves called-upon to emotionally support logical and biased partners, agitated by unsound reasoning. That's not to imply that women would've chosen that path – they wouldn't, in-so-far as the path itself was undeniably oppressive. Nevertheless, as if taking inspiration from that contentious axial refinement, telepathically-induced effects ought to be applied in a structured manner, not unlike the **functionally-nucleated family**. A functionally-nucleated approach which betrays sapient deficiencies, whilst coevolving aequipine behaviours in both sexes.

- Genuine expressions of affection

Both the mind and coevolution advance cyclically, enabling one to rewind the anthropological clock, thereby facilitating retrodictions regarding humanity's past. By the same token, one could also cyclically extrapolate, winding things forward at an accelerated pace, prior to returning to the present in order to avert disaster.[31] Thus, cruciform calculus enables humankind to prophesize about the past, present and future using a dataset available only to an **epistemic community** of balanced individuals, every one of whom is acquainted with telepathically-induced effects. **Post-radical feminism** concedes that female aequiponimity owes much to domestic subjugation – hence some women 'subordinating' themselves in order to retain their axial advantage. Begging the question, how far should society go in replicating that controversial achievement?

Feminists like Christine Delphy (1941–) and Diana Leonard (1941-2010) saw the family as an unfair hierarchical institution – one in which the bread-winning patriarch felt compelled to express 'love', even as the burden of unpaid domestic labour played-out asymmetrically. However, that selfless intra-family concern has left an indelible mark, with women's brains becoming less lateralized than men's over the course of time. The question is, what must be done to men to replicate that achievement? As luck would have it, male subordination creates a constructive burden, but scarcely one more onerous than domestic servitude. With that in mind, it's much too early to start generalizing about the family's merits, its failings reflecting upon the genetic substructure of contemporary society.

Clearly, the coevolutionary dynamic pervading that nucleated and socialized domestic structure isn't yet finished, such that feminists might want to reserve judgement on the functionally-nucleated family unit until nearer to the anthropocene's end. In the meantime, the proffered neocentric 'race to the top', which this book's cruciform strategy actively promotes, could see groups and individuals vying for the telepathic advantages conferred on them from above – in the first instance, by proving themselves balanced-minded. And, whilst it's difficult to objectively judge a normative structure, provided the genus *Homo* develops an aequipine genome, many of society's templates should serve humanity just fine – with cruciform advance making love, sex and gender the emotional foundations of a rewarding coevolution.

- Emotional triggers

Those attention-grabbing pictures, published daily by the **news** and social media, provoke anger, fear, hope, despair, love and hate. One such image was that of the confederate battle flag being removed from the South Carolina Capitol Building, USA, back in 2015. Prompting former US President, Barack Obama (1961–), to tweet, that its removal was "*a signal of good will and healing, and a meaningful step towards a better future*". President Obama was well-placed to judge, having signed into law the Matthew Shepard and James Byrd Jr Hate Crimes Prevention Act, in 2009. Matthew Shepard (1976–98) having been cruelly murdered by homophobes (when found, his head bloodied except for the tracks of his tears). As for the ill-fated African-American James Byrd Jr (1949–98), he was dragged to his death behind a pick-up truck by white supremacists sporting racist tattoos, including the confederate battle flag motif.

More **sub-aequipine behaviour** (linked to the proto-fascist '*stars and bars*' and its sapient derivatives) surfaced in 2015, when a self-professed white supremacist committed the Charleston Church Massacre, having posted images of himself online brandishing that psychologically-disturbed emblem of concentrated white power. Given the enormity of such crimes, it takes a considerable amount of emotional restraint, allied to Cartesian levels of reasoning, for the '*stars and stripes*' to fairly reflect aequiponimity. Seen from a suitably detached perspective, the subordination of mainstream agents takes place without explicit reference to high-profile happenings – a move predicated on the principle that by displacing bias and logic, one inevitably undermines the biological roots of xenophobia, supremacism and hate crime.

Professor Alex Pentland (1952–), who's built a career studying digital social networks, has argued that social learning, via peer-to-peer communication, is what actually changes human behaviour (as opposed to received learning, undertaken in a formal manner).[32] As a consequence, all that liberal tutoring – as regards nationalism, racism and sexism – impacts less upon personal behaviour, than the ill-judged social fabric which exists between fraternizing peers. Needless to say, peer-to-peer communication encourages **dumbing-down**, and a perceptible "*the law is whatever you can get away with, irrespective of the law*" attitude. More broadly, there's nothing civil about 'civil society' corrupting textbook protocols (with peer-to-peer exchanges prompting poor cordite handling at the Battle of Jutland, in 1916, the impact of which killed hundreds of British seamen).[33]

For convenience, we can say that peer-to-peer communication and social learning are aspects of the **social action crucible**. That **crucible of social action** being driven, for the most part, by heightened emotion, rather than cool detachment. However, whilst subordinate peer-to-peer exchanges lead to spiralling sapiency, comparable exchanges between supraordinate individuals lead to spiralling aequiponimity (amounting, in most cases, to individuals honouring the recognized texts and established protocols). Thus, the sub-aequipine racist hate passing between James Byrd Jr's assailants tipped the balance in the direction of murder, whilst the horror of that crime energized its sound-minded investigators, who conspired to honour due process. *Homo sapiens*, you see, is more 'endocrinal' than 'cognitive' – its antipathetic generalization bearing-down unspeakably upon the minorities, as their pecuniary and prestige small talk escalates alarmingly.

- The universe's primary purpose

To paraphrase German mathematician Emmy Noether (1882-1935), "*for every conservation law, there's a symmetry*". A symmetry arising whenever a system undergoes a quantifiable change, but aspects of it remain unaltered. If the universe could be said to have a purpose, it's a reduction in destabilizing asymmetries. Thus, cruciform laws permit the political picture to alter alarmingly, whilst justice is conserved long-term. If an unsound emotion is an asymmetry, then permitting those feelings to cloud the truth becomes an astronomical error. Such sapient emotionality driving forward the wider concentration of power, as evidenced by the horrors facing those who attempt to decompose the same. With aequipines focusing on the truth, from which feelings habitually arise, and **sapients** obscuring the truth, due to misplaced sentiments.

Who knows, perhaps the upward-pointing finger of renaissance art is demonstrative of a slow-progression from asymmetrical bias and logic towards an enlightened symmetry founded upon sound and balanced thinking. And, whilst the playwright Somerset Maugham (1874-1965) sardonically proposed that love was a dirty trick played on all humanity in order to achieve the continuation of the species, less irreverent individuals would interpret it as a phenomenon which '*the heavens*' are assiduously working towards conserving. Assuming that '*the heavens*' aren't wrong, then the aforementioned 'race to the top' concludes with love – love being central to conceptual reality's axial advantage of choice.[34]

Presumably cruciform coevolution has its very own crossover day – no, I don't mean the day when sufficient Brexiteers have died, and enough remainers have become eligible to vote, that the UK has once again become irrefutably Europhile – but rather, that point in time when aequiponimity has become the *de facto* cognitive, nervous and endocrinal response. That axial advantage reflecting a supraordinate's real time processing of sensory and extrasensory information – aequipines having been well-served by their radiological, electrochemical and hormonal reactions, and sapients having been fatally weakened by the same. Either way, that chronological haze, which we euphemistically term time, blinds us to events beyond our narrow comprehension – with axial differences becoming most apparent, whenever our survival is contingent upon penetrating that particular miasma.

Ch14: **Sensory deprivation**

- Devastating asymmetries

Sensory deprivation leaves palpably avoidable tragedies in its wake. But, to what extent should technology be used to compensate for subordinate reasoning? Having eliminated circumstance as the probable cause of a tragedy, all that remains are idiosyncratic cognitive, nervous and endocrinal functions. With that in mind, the 27th March 1977 would've progressed very differently had those involved in the Tenerife Air Disaster been adequately vetted for the balance of their minds and the soundness of their judgements. **Soundness** being as much of a commitment to verifying the facts, as it is a reflection of one's ability to deal with a scarcity of information. Therefore, when Pan American flight 1736 (a Boeing 747, *en route* from Los Angeles to Las Palmas) found itself fog-bound on the solitary runway of the highly-congested Los Rodeos Airport, having previously been diverted, sound awareness became vital.

With fog rolling-in, and no ground radar available, confusion quickly arose amongst the queuing pilots. Nevertheless, air traffic control (ATC) advised the Pan Am plane to "*taxi down the runway, and – ah – leave the runway third, third to your left*" [background noise, associated with a soccer match playing on a radio, was later said to have distracted the controllers]. Pan Am responded "*third to the left, okay*". ATC then added "*third one to your left*". After some deliberation, Pan Am asked "*would you confirm that you want the clipper 1736 to turn left at the third intersection?*" Whereupon, ATC responded "*the third one, sir; one, two, three, third, third one*". To which Pan Am responded "*very good, thank you*". Then, ATC ambiguously added "*. . . er, 7136 (sic) report leaving the runway*" (which, in the growing confusion, could have been interpreted as them *having* left the runway).[35]

At this point, the Pan American aircraft – and another jumbo jet, KLM flight 4805, belonging to the Dutch national airline – were told that the centerline lights were "*out of service*". The Pan Am crew associated this information with the stipulation that departing aircraft must have a minimum of 800 metres visibility, making an immediate take-off appear impermissible. If the Pan Am flight thought taking-off implausible and the KLM flight, waiting for clearance to take-off, was equally convinced that the Pan Am flight had taxied-off the runway, both parties would be edging towards disaster. In fact, the only thing dissuading Jacob van Zanten

(1927–77), the KLM's captain, from taking-off, was the issue of formal clearance, as he told his flight engineer that the Pan Am flight *was* clear, when questioned. And so, when van Zanten boldly announced *"let's go, check thrust"*, he was convinced he was acting prudently (ATC replying "okay", to the comment *"we're now at take-off"*).

In the unprecedented carnage which followed, van Zanten's fuel-laden Boeing 747 struck the Pan American aircraft with such force that both planes violently disintegrated, their aviation fuel igniting into a deadly fireball, killing 583 people (and leaving just 54 survivors). The **third measure of soundness**, in addition to determining the facts and managing a deficit of information, is appreciating how information is received by third parties. As the ATC failed on all three counts, responsibility for the tragedy must lie with the Spanish air traffic control. After all, the Pan Am crew's confidence that the KLM flight wouldn't take-off owed less to formal clearance, and more to the protocols surrounding visibility. On reflection, the KLM's flight engineer was the soundest person present, but their doubts were summarily dismissed by the plane's captain (that is to say, Abelard's dictum might still have saved them).

Although ground radar was subsequently installed at Los Rodeos Airport (later renamed Tenerife North Airport), the concern is that technological advance is simply obscuring humanity's growing capacity for sub-aequipine human error. Ultimately, there will always be heart-stopping moments when the brain's **frontal lobes** shut down, and pure instinct takes over. The question is, are those radiological, electrochemical and hormonal instincts sound or unsound? For example, would keeping the KLM aircraft on the ground have saved lives – effectively ramming the other aircraft? Certainly van Zanten didn't think so, which is why he chose to gouge an almighty hole in the runway with the plane's tail section, having placed his weighty fuel-laden aircraft into a hopelessly deep stall. Chillingly, the healthy mind paradox guarantees that momentary beliefs are positively-reinforced, making otherwise preventable catastrophes appear practically unavoidable.[36]

- Fourth measure of soundness

The Dutch authorities, in response to the Spanish-led Tenerife Air Disaster Inquiry's conclusions, stated that there was no disagreement over the established facts, only the manner in which those facts were interpreted. Indeed, the Dutch were quick to observe that given the density of the fog, the KLM captain's

Figure 25: Neurotypic studies (principal academic fields)

	neologism	Greek	meaning
aequipine	ameroliptics	*amerolipsia*	impartiality
aequipine	epivevatics	*epivevaiosi*	confirmation
sapient	hermeneutics	*ermineia*	interpretation
sapient	epipoliotics	*epipolaiotis*	cursoriness

The above table lists the principal academic fields tasked with exploring balanced, sound, logical and biased (that is to say, the origin, impact and expression of such reasoning) absolute conviction that they were safe to take-off could only have been arrived at through radio communications. *Ergo*, it was the sub-optimal functioning of the Spanish air traffic control which was responsible for the catastrophe. Accordingly, the **fourth measure of soundness** denotes one's ability to correctly interpret the facts – with the Spanish failing to determine the pertinent facts, failing to manage a deficit of information, failing to appreciate the impact of their communications on third parties and failing to interpret the evidence correctly. Given the magnitude of those failings, technology would appear to be their only hope.

Many so-called professionals are governed by what they anticipate 'getting away with', leading to sloppy government, lackluster parliamentarians and trigger-happy police. Such individuals interpret the permissible speed limit as '*any velocity which isn't perceptibly policed*', the Geneva Convention as '*any military deed which isn't conspicuously challenged*', environmental obligations as '*those acts which don't yet incur a penalty*', and political representation as '*the bare minimum which must be done to remain in office*'. Brexit, for example, may result in liberties being taken with established protocols, making provincial autonomy appear redolent of an **autistic condition** requiring third-party support. Proof, as we enter the 'mist and fog' of escalating trade wars, that simply possessing social skills is no guarantee of heightened cognitive ability.

In practice, matters of interpretation are 'axially' resolved by means of **radiotransmission**, **neurotransmission** and **arteriotransmission** (with the least detached individuals tendering the most emotive, or crudely basal, sub-aequipine explanations). Indeed, those interpretations often serve to highlight who is '*less than logical*', '*logical*', '*sound*' or '*more than sound*', making the subject positively Linnaean in its utility. As one's heteronnubial co-author shrewdly asserts, the driving force behind each and every interpretation is perception – perception combining *bottom-up* modality-driven processes, related to the five senses, with *top-down* conceptual processing, drawing heavily on what has been taught or experienced. However, whilst aequipines cast doubt upon desultory assessments, sub-aequipines positively-reinforce the same, leading to deepening subjectivity.

• Rewarding subordination

The scientific fields responsible for examining biased, logical, sound and balanced (as regards their origin, impact and expression) are respectively termed **epipoliotics** (from *epipolaiotis*: Greek for cursoriness); **hermeneutics** (from *ermineia*: Greek for interpretation); **epivevatics** (from *epivevaiosi*: Greek for confirmation); and **ameroliptics** (from *amerolipsia*: Greek for impartiality).[37] Before semiotics, we experienced anger, fear, love, despair and hate as biologically-produced basal responses. Today, thanks to our symbolic socializing processes, we're able to socially-construct emotions (a person's disposition reflecting both basal responses and emotions). And so, depending on one's **neurotypic sub-classification**, one may experience epipoliotical, hermeneutical, epivevatical or ameroliptical emotions – *Homo sapiens'* emotions being compromised by basal responses and *Homo Aequipondium's* basal responses being tempered by erudite emotions (see Fig. 25).

These scholastic fields help us to understand what must be measured, in order to categorize individuals. Moreover, they help us to determine whether a person is hindered by their own biology or by irresponsible social conditioning. Objectivity, neutrality and detachment increase as one ascends through epipoliotics, hermeneutics, epivevatics and ameroliptics; just as partiality, subjectivity and imagination decrease as one ascends through the same (that is, the subjects themselves, not the clinicians studying the same). It's important to appreciate that not everyone classified as subordinate is objectionable, it simply means that they're ill-suited to maintaining the structures providing for telepathically-induced effects. For its part, art can be primitively basal, passionately emotional or intellectually contrived – with some sound-minded artists abandoning objective truth altogether, in favour of an unwavering bias towards an *avant-garde* style.

But how does one categorize the purportedly unsound, participating in a 'race to the bottom'? In many respects, it's the responsibility of those academics, working in the fields listed above, to ascertain the respective merits of those committed to the same. What's clear, is that mawkish, inoffensive and literal art works are derided as 'chocolate box' or 'naff', making creative rebellion – even that attracting remotely-induced effects – appear more appealing. Of course, it's terribly 'clever' and pretentious to feign subjectivity, especially if you happen to be objective – but, charlatans aside, subordinate psychology offers an unusual perspective. And so, creative spirits may triumph through the patronage of balanced minds – with aequiponimity expressing the **objective truth** and sapiency a much vaunted **subjective truth** (with the latter attracting astronomical prices, notwithstanding that the former is more priceless).

•   Obedience to knowledge

Mark Twain (1835-1910) once wryly observed that the two most interesting characters in the 19th century were Napoleon (1769-1821) and Helen Keller (1880-1968). Helen having been born sighted, with normal hearing, but having become deaf and blind at the age of 19 months due to an unknown illness. A highly intelligent child, she became increasingly frustrated by her inability to converse, resulting in disorderly behaviour. Until, that is, Anne Sullivan (1866-1936) became her teacher and companion. Sullivan, herself visually-impaired from a young age, was a former student of the Perkins Institute for the Blind (whose director specifically chose her for that role). It was Anne who helped Helen to decode the "*living world*" of language through finger-spelling, sufficient for her to converse with other children. Helen later recalled that it "*awakened her soul; gave it light, hope, joy, ... set it free! Everything had a name, and each name gave birth to a new thought*".

"*Oh, what happiness*", exclaimed Keller, at her ability to construct a self-schema, converse with friends and learn about the world – eventually becoming the first deaf-blind person to achieve a Bachelor of Arts degree. So many positive emotions *are* socially-constructed that one cannot overstate the importance of semiotics, as regards experiencing first-hand such things as elation, exhilaration and euphoria. In the event, Helen, who benefited enormously from having a balanced persona, became more of a person than many of today's chemically-dependent juveniles, foolishly shunning a decent education. Anne herself declared, "*only when we have worked purposely and long on a problem, ... and, in hope and despair wrestled with it, ..in silence and alone, relying on our own unshaken will, ... only then, have we achieved education*".

Keller understood implicitly the value of symbolic socializing processes, as regards building and sustaining a positive mental attitude. She accomplished so much because she collated data perceptively, managed a deficit of information with genius and appreciated the impact of her communication on others. Moreover, she interpreted the facts with postgraduate maturity and understood the limits of her knowledge. In many respects, she humbled those possessing all of their **modalities**, and taught us much about the construction of personality. She was, at heart, a person with severe disabilities, who risked becoming a severely disabled person.[38] On reflection, the emotions which she socially-constructed, in both herself and others, remain some of the most positive ever fabricated – making her, without question, a far more interesting character than Napoleon.

- ### How the brain works

The foetal brain experiences hypertrophy (an increase in the size of neurons), hyperplasia (an increase in the number of neurons), differentiation (the specialization of neurons) and structural refinement (whereby comparable neurons organize themselves into functionally-specific systems and bodies). The **depolarization** of contiguous brain cells is an energy-demanding process, powered by **ATP hydrolysis**. Although powered by adenosine triphosphate, those nerve impulses are *triggered* by either adjacent neurons or absorbed thoughts. As a result, the instigation of this **absorption-impulse** phase is endergonic, requiring energy from thoughts, brain waves or neurotransmitters. Conversely, the **inactivation-emission** phase is exergonic, with energy lost through the expulsion of radio waves or the release of neurotransmitters at the synapse, thereby facilitating or inhibiting the passage of the signal.

Figure 26: Neuronal induction (the neuron as a logic gate)

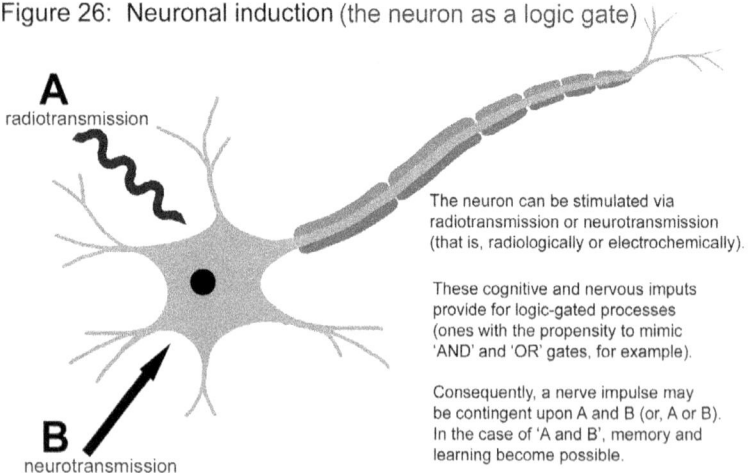

A radiotransmission

B neurotransmission

The neuron can be stimulated via radiotransmission or neurotransmission (that is, radiologically or electrochemically).

These cognitive and nervous imputs provide for logic-gated processes (ones with the propensity to mimic 'AND' and 'OR' gates, for example).

Consequently, a nerve impulse may be contingent upon A and B (or, A or B). In the case of 'A and B', memory and learning become possible.

Leaving aside the influence of **hormones**, the human brain constitutes an electrochemical and radiological hub, whose person-specific absorption profile reflects the spectroscopic impact of neuromagnetism. As each neural pathway is laid-down the brain is showered with radio waves, triggering further signals. Once begun, this runaway cyclical process eventually reaches such staggering sophistication that the radiological dimension begins to surpass both the electrochemical and hormonal alternatives. At that critical moment, conscious self-awareness takes over and the placental mammal – sorry, human being – is able to socialize, thanks to syntax, semantics and pragmatics. Replicated digitally, the computer would become conscious of its own existence – that is to say, logic-gated computers and synaptic brains both mediate between inputs and outputs, using untold Boolean operators (see Fig. 26).

By making a neuron receptive to a specific frequency, neuromagnetism guarantees that a brain cell is either actioned or it isn't (depolarization, due to adjacent cells, notwithstanding). Which, in **binary notation**, translates as two alternate states, comprising either a 'one' or a 'zero'. Absorption and emission, at the molecular level, barrages the brain's integrated circuitry with binary information – with QSD and QED focusing respectively upon the absorption and emission of thoughts, in a process termed **neuronal induction**. As an emitted thought can possess an entirely different wavelength to an absorbed thought, neurons may activate sequentially. The true significance of brain cells 'firing', therefore, may lie in the radiological ramifications of the same, rather than the electrochemical alternatives. Thus, the human brain's frenetic cycle of development is concomitant with the conceptual's subjugation of the corporeal.

Given humanity's predilection for mainstream duplicity, self-deception and official denial, it may take a '*divinity*' to competitively select the most pertinent forms of inculcation, the most scientific of interpretations and the most gifted of epistemic communities. In other words, we may find that humankind's efforts are being '*judged*' by conceptual forces far exceeding *Hominini's* feeble binary processes. Herein lies the discomforting truth, as regards the human condition. Namely, that our brain's Boolean mechanisms tempt us with the prospect of knowledge, whilst actually highlighting a broader multisensory impairment. Which is why epivevatics and ameroliptics take us beyond the available evidence, to issues such as incomprehension, misapprehension and imperceptibility. That is to say, by doubting ourselves, we come, not to humanism, but to something astronomically more structured, symmetrical and conserved.

- Nietzschean perspectives

Friedrich Nietzsche (1844-1900) argued that the truth was unattainable – believing, instead, that there are only perspectives or interpretations, driven by the "*will to power*". Nietzsche was, to all intent and purposes, an **extreme sceptic**, for whom the phrase '*cogito, ergo sum*' might easily have been coined. He viewed certainty as unachievable, values as objectively baseless and life as completely without purpose. Like later feminist writers, Nietzsche was, without consciously realizing it, addressing the deficiencies and contradictions inherent in *Homo sapiens*. In that sense, his proffered **nihilism** reflects, not-so-much the human condition, but rather, the sapient condition – with *Homo sapiens'* weakness for **Nietzschean crises** made all the more palpable, by virtue of its piecemeal unsound thinking and long-term sub-aequipine behaviours.

After all, certainty *is* unachievable, values objectively baseless and life utterly without purpose if one is governed by biased and logical analysis. But, the same cannot be said of *Homo aequipondium*, which benefits from sound and balanced cogitation, making it anything but nihilistic. Place *Homo aequipondium* and *Homo sapiens* together in a finite space, with only limited resources, and the former will out-compete the latter, but not without the latter negatively-impacting upon the former (an extirpating process made all-the-more efficient by virtue of *Homo sapiens'* time-worn structural perspectives being either inherently unsound, or else finding themselves ruinously weakened by sub-aequipine elements). In that sense, cruciform coevolution, in spite of its many Nietzschean crises, remains very much the social system of choice.

Clearly, some 'axes' are more competitive than others, hence the proffered disciplines of epipoliotics, hermeneutics, epivevatics and ameroliptics. Armed with these neurotypic sub-classifications, one can evaluate structurating agents and socializing structures, sufficient to foretell the actions of the proverbial 'seamen', relative to the fate of the hypothetical 'dreadnought'. What's clear, is that those who exaggerate the enormity of known facts, whilst grave unknowns threaten some terrible misfortune, are manifestly neither sound nor balanced. For example, the Right-wing scapegoats to appear tough – its sapient hardliners, reduced to bungling by bias and logic, routinely overstating known misdemeanors in the hope of appearing superficially effective. In that sense, the sapient condition is one of looming catastrophe – much of it self-inflicted.

Ch15: **Basal responses**

• One axis, two cycles, three systems

The brain cell is, to all intent and purposes, mother nature's very own **transistor**, with the dendrites serving as the collector, the axon as the emitter and neuronal induction as the base. In electronics, a weak current in the transistor's base regulates a much larger flow of electricity from the collector to the emitter – with neuronal induction, i.e. thoughts and brain waves, hypothesized to regulate the flow of nerve impulses within the brain. Accordingly, the human brain comprises countless logic gates – be it neurons, **nuclei** or **ganglia** – all of which rely on neuronal induction to computationally-integrate their many electrochemical functions. Not only does neuronal induction coordinate the flow of **electrochemical signals** within the brain, the resultant activity is monitorable by means of **brain waves** emitted during the neuron's **absolute refractory period**, i.e. the brief rest period, during which the brain cell can't be re-actioned.

This radiological synchronization of the brain's countless cognitive functions involves an astonishingly light touch – requiring, as it does, seemingly irrelevant amounts of emitted energy. According to orthodox medicine, the human body possesses two major **communication systems**: one nervous (**electrochemical**), the other endocrinal (**hormonal**). This book, as outlined earlier, introduces a third: namely, the cognitive (**radiological**). This triumvirate of co-dependent neurophysiological systems continually interact, sufficient to produce *'emergent behaviour'*, *'emotional criticality'* and *'acute non-linear outbursts'* – the electric and magnetic fields imperfectly inscribing, with the help of **hormones**, that *'uniquely-blank, person-specific material'* known simply as one's brain.

Consciousness arises due to the mind cycle transforming the brain from a predominantly corporeal state, into a primarily conceptual one. The **cognitive system** ceasing to be a by-product of the **nervous system** – with cogitation becoming, instead, the principal cause of that consciousness-sustaining electrochemistry. As for the **endocrine system**, that's managed by the nervous system in a largely, but not exclusively, involuntary manner. A person's hormonal responses betraying the mind's sentient control over the brain's integrated circuitry. Thus, punishing and rewarding radiotransmission, neurotransmission and arteriotransmission

143

heighten semiotic assimilation in that sensitive period preceding conscious self-awareness, making early years learning comparatively visceral.

A person's disposition is the product of one axis, two cycles and three systems, formed from specialist eukaryotic cells. Those specialist eukaryotic cells, like electrons, come with certain preconditions attached. However, whilst the electron moves in response to fluctuating conditions, the specialist cell must have its conditions maintained through homeostasis. Consequently, the **endocrine** system releases chemical messengers into the bloodstream, so as to help the brain and body function optimally. The mind intruding whenever a person encounters, or anticipates, stress – whereupon these radiological, electrochemical and hormonal effects cascade, generating enhanced blood-flow, deeper breathing and an escalating heart-rate. In other words, the **prefrontal cortex** may stimulate the body's subsidiary **hypothalamic-pituitary-adrenal axis** (HPA-axis).

Figure 27: CNE-axis (comprising two divergent cycles)

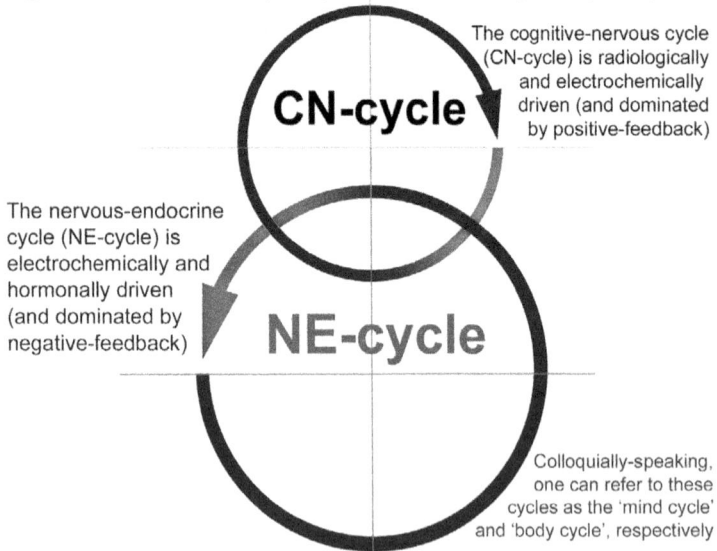

CN-cycle

The cognitive-nervous cycle (CN-cycle) is radiologically and electrochemically driven (and dominated by positive-feedback)

The nervous-endocrine cycle (NE-cycle) is electrochemically and hormonally driven (and dominated by negative-feedback)

NE-cycle

Colloquially-speaking, one can refer to these cycles as the 'mind cycle' and 'body cycle', respectively

The human brain's hypothalamus, which is located within the **diencephalon**, integrates both the nervous and endocrine systems, by way of the pituitary gland, in an archetypal case of negative-feedback.[39] In short, the hypothalamus, in response to low cortisol levels, secretes corticotropin-releasing hormone (CRH), which tells the pituitary gland to release adrenocorticotropic hormone (ACTH), which instructs the adrenal glands to boost cortisol levels (until further CRH secretion is inhibited). Needlessly triggering this highly-developed stress response has the effect of disrupting diurnal biorhythms, suppressing appetite, increasing gastric-acid production and weakening immune responses – which is why unsound bosses ought to be penalized for making impossible demands. Whilst highly-sophisticated, the HPA-axis is nonetheless eclipsed by the all-encompassing cognitive-nervous-endocrine axis.

What makes the cognitive-nervous-endocrine axis (CNE-axis) so fascinating is that it comprises two extraordinarily divergent cycles: 1) the radiologically and electrochemically driven **cognitive-nervous cycle** (**CN-cycle**), which is dominated by neurotransmitters and positive-feedback; and 2) the electrochemically and hormonally driven **nervous-endocrine cycle** (**NE-cycle**), which is dominated by negative-feedback and hormones (see Fig. 27). Strictly-speaking, the mind cycle is a colloquial term for the CN-cycle, with endocrinology the academic study of the NE-cycle. In practice, there are moments of sheer terror when the CN-cycle is hostage to the NE-cycle, and other occasions when the NE-cycle is wantonly actioned by the CN-cycle. Either way, an **ameroliptical** outlook is a coevolutionary must-have, arrived at through the piecemeal fine-tuning of these seemingly incongruent cycles.

• Pure positive-feedback

Human beings are 'axes', centred upon cognitive, nervous and endocrinal systems. Crucially, the nervous component serves two contradictory cycles, using remarkably similar methods – namely, chemical messengers such as monoamines, **amino acids** and peptides. The **triune brain**, an idea first proposed by American neuroscientist Paul MacLean (1913-2007), depicts placental mammals' brains, and hence those of humans, as comprising a reptilian complex, a paleomammalian complex (or **limbic system**) and a neomammalian complex. That paleomammalian complex, where it meets the **basal ganglia** (or reptilian complex), being the focus of the writers' proposed CNE-axis (as described in the glossary entry for **brain**). So, whilst endocrinal activity constitutes a lesson in negative-feedback, replete with symmetry and conservation, the cognitive system appears ridden with positive-feedback and destructive asymmetries.[40]

Significantly, the mind derives from a chain-reaction of thoughts, impulses and exchanges – many with the capacity to exterminate the owner. Such runaway 'neural fission' constitutes a powerhouse of cyclical possibilities, hence evolution's longstanding attempts at diminishing the mind cycle's speed, function and yield. Accordingly, at the heart of human consciousness lies something biologically frenetic. So harmful, in fact, that mother nature has given us **apoptosis**, or programmed cell death, whereby roughly half of all brain cells are rendered defunct. The healthy mind paradox arises because, in spite of the human body possessing clearly defined optimal conditions, cognition must, by its very nature, possess disturbing amounts of plasticity (notwithstanding, that ameroliptical and epivevatical **neuroplasticities** rate more highly than hermeneutical and epipoliotical ones).

That primordial transition from pre-biotic to biotic was driven by the need to self-regulate around a range of rapidly-evolving physiological set-points. First-and-foremost, by harnessing ATP in order to pump ions back-and-forth across a specialist cell's plasma membrane (such active transport, in pursuit of electroneutrality, amounting to negative-feedback). But what of the CNE-axes' exploitation of positive-feedback? That began with **bioelectrogenesis**, whereby a cell's chemical energy was used to *expel* ions from inside the cell, sufficient to polarize its plasma membrane (the cell subsequently depolarizing when 'fired'). This being the first step towards developing a fully-fledged nervous system. Significantly, the eukaryotic cell had discovered positive-feedback – a revelation made all-the-more earth-shattering by virtue of the cell's absorption and emission of energy.

What's more, chemical messengers began to integrate this manic neurological network, via the ubiquitous synapse, amid burgeoning apoptosis (with comparable chemicals communicating with the body, owing to the actions of the endocrine glands). Thus, the neuron became the *de facto* key component within the emergent brain's rapidly-expanding circuitry. An electronic component which, in addition to being a logic gate, or 'transistor', could also amalgamate into nuclei or ganglia, in order to process signals in a predetermined computational manner. But could so much electrically-frenetic organic matter sustain a near-transcendental radiofrequency dimension? With an anthropic inevitability, conceptual ascendency amid corporeal submission led to human consciousness – and with it an outright ethereal detachment, in-keeping with a '*heaven-sent*' domination of material reality.

- Every bit of one's soul

Each region of the brain comprises specialist neurons, custom-built to process signals conveyed to them by means of radiotransmission, neurotransmission and arteriotransmission. The CN-cycle's computational 'bit', as regards the binary processing of information, is the **unstable multivibrator**. The unstable multivibrator being the hypothesized building-block of human cognition, with *"every bit of one's soul"* comprising a unique wavelength. At its simplest, the unstable multivibrator consists of two neurons ('A' and 'B'), which are able to excite one another in an inexhaustible manner (see Fig. 28). Adjacent neurons ('C' and 'D') serve as the **axonal input** and **dendrital output**. Because the absorption-impulse and inactivation-emission phases of 'A' and 'B' are staggered, radiotransmission is able to sustain uninterrupted neurotransmission across their synapses.

Figure 28: Unstable multivibrator (frequency modulation)

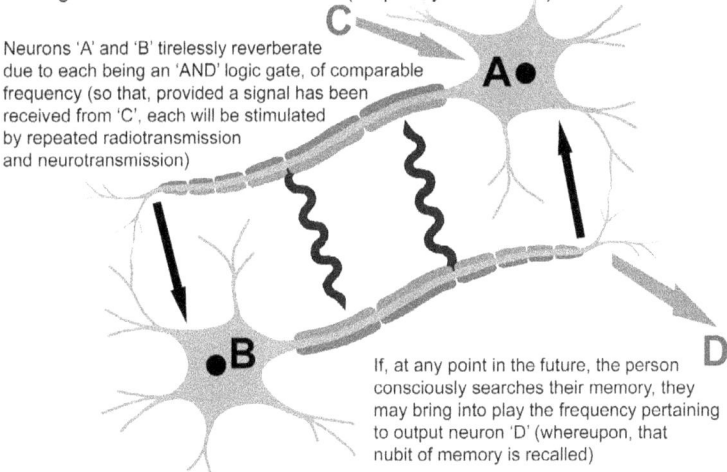

Neurons 'A' and 'B' tirelessly reverberate due to each being an 'AND' logic gate, of comparable frequency (so that, provided a signal has been received from 'C', each will be stimulated by repeated radiotransmission and neurotransmission)

If, at any point in the future, the person consciously searches their memory, they may bring into play the frequency pertaining to output neuron 'D' (whereupon, that nubit of memory is recalled)

In other words, once a signal has been received (via axonal input 'C'), the unstable multivibrator will tirelessly reverberate, due to neuronal induction. Assuming that the axonal input is then deactivated by apoptosis, the multivibrator will then exist as a 'bit' of one's **memory**. As each quantum of memory has its own wavelength, the mind is **frequency modulated**. Moreover, apoptosis doesn't impose upon memories, it makes them! According to this synopsis, the human mind frequency modulates every irreducible piece of information that it memorizes, sufficient to engage in *top-down* conceptual processing. It's important to note that each 'bit' comprises a one or a zero – in other words, the multivibrator may not reverberate at all. To access the same, neuron 'D' must be stimulated with a completely different frequency, whereupon the mind can establish whether a one or a zero has been 'saved'.

Hypothetically, the human mind is able to consciously probe the recesses of its own memory, due to complex layers of radiological organization. Therefore, neurotypic sub-classifications reflect: 1) the manner in which one's modulated memory is radiologically accessed or read, which could involve semi-permanent brain waves or passing thoughts, plus the soundness of that inculcation; 2) the corporeal structure of the brain, including the effectiveness of its neural pathways and neurotransmitters; and 3) the psychological impact of hormones, in terms of their neurophysiological functions, relative concentrations and homeostatic effects. Collectively, the clinical fields termed epipoliotics, hermeneutics, epivevatics and ameroliptics deconstruct each of these key areas, in order to better understand biased, logical, sound and balanced dispositions.

Chemical messengers, e.g. neurotransmitters (such as dopamine, serotonin, oxytocin and endorphins), hormones (such as cortisol, adrenaline, insulin and glucagon), and the male and female sex hormones, routinely fluctuate, thereby affecting a given person's physical, mental and emotional wellbeing. However, the principle reason why psychiatry, with its emphasis on medication, and psychology, with its commitment to psychotherapy, struggle, is because the human mind derives from radiofrequency events transcending both nervous and endocrinal activity. Therefore, it's easy to influence one's mental state using drugs and alcohol, but incomparably harder to undo years of defective radiological organization. Indeed, having organized brain waves into a spirited juvenile psyche, the adolescent mind then possesses emergent properties stubbornly resistant to revision. All of which begs the question, what is clinically normal?

• Clinical normality

Sports psychology, at its simplest, amounts to vigorously warming-up, whereupon raised endorphin levels leave one feeling self-assured, indefatigable and confident. This 'runners high' often proves addictive, with inactivity triggering symptoms akin to endorphin withdrawal, e.g. guilt, anxiety and depression. Oxytocin, another 'feel good' neurotransmitter, is commonly associated with pair-bonding, intimacy and trust, making it connubial dualism's most potent biochemical ally. Dopamine, on the other hand, is exceptionally motivational, due to its mood-enhancing characteristics. But, as the habitual drug-addict soon discovers, this chemically-induced 'high' rapidly spirals into tolerance and dependency. As for serotonin, that supports normal physiological function, but is lethal in excess; forcing the concerned sceptic to ask, "*is serotonin syndrome a smokescreen for disorders arising from the misapplication of radiofrequency technologies?*"

Using neurotransmission, the autonomic nervous system's **sympathetic** and **parasympathetic** branches 'excite' and 'inhibit' a panoply of bodily functions, sufficient to compliment antagonistic hormones such as glucagon and insulin, the primary regulators of blood glucose levels. In the case of the sympathetic nervous system, the body's **sympatho-adrenal stress response** raises adrenaline concentrations in the blood – such 'excitation' boosting cortisol secretion, which compliments glucagon by raising blood glucose levels. Simultaneously, that sudden rush of adrenaline increases one's heart-rate, deepens one's breathing and dilates one's pupils (making adrenaline antagonistic to insulin and pupillary dilation a symptom of distress in those with diabetes). Whether remotely-induced radiotransmission could save the diabetic's life is an open question. However, one's own insulin normally restores blood glucose levels, via life-preserving negative-feedback.

If one defines **clinical normality** as 'a drug-free physiological state, not requiring pharmaceutical correction', then legalizing illicit drug-use becomes an improper normalization of abnormal cravings. What's more, according to this highly-exacting definition of clinical normality, any remaining drug use becomes evidence of medical necessity, mental illness or an unhealthy lifestyle. Recreational or habitual drug use often arises in youth prior to the advent of palpable stress. When unavoidable stress does eventually materialize, those earlier bad habits turn into unshakable dependencies. Judged from a political standpoint, any welfare system which subsidizes the chemical dependencies of young couples, over helping single homeless people, warrants grassroots reform. After all, only *Homo sapiens* would perceive self-inflicted chemical dependency as healthy, normal and desirable.

- Multivibrational matrices

The unstable multivibrator, otherwise known as the **neurological bit** (or nubit of memory), is sustained by a specific wavelength. Consequently, the nubit is able to output a nerve impulse provided the outgoing signal is impelled by a brain wave, or conscious thought, of differing frequency. In the absence of such a brain wave, or conscious thought, the nubit of memory simply languishes – and, in time, might even disappear. Therefore, memory exists at the bottom of a complex hierarchical assemblage – with the *top-down* conceptual processing of information taking place within the brain's outermost neocortex, while memories are formulated deep down within its limbic system. Any contemporary image of the brain which you see is sorely lacking, in-so-far as the cognitive system exists as a frequency modulated matrix, ghosting in-and-around the brain's physical structure.

Figure 29: Multivibrational matrix

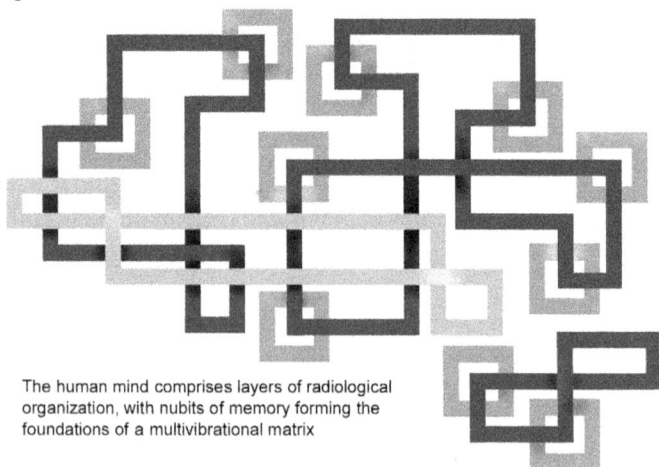

The human mind comprises layers of radiological organization, with nubits of memory forming the foundations of a multivibrational matrix

Using **multivibrational psychometrics** one can generate a schematic map of the cognitive system, sufficient to navigate the mind's labyrinthine pathways, quantized memory and parallel processing (see Fig. 29). One's mind mimics the electronic configuration, in-so-far as it concentrates electromagnetic energy within a structure possessing multiple layers. Moreover, those layers, or cognitive levels, govern not only the accessing of memory, but also the progress of electrical currents. At the conscious level, thoughts are fleeting. However, one's deeper subconscious, and impregnable unconscious, comprise layer upon layer of semi-permanent brain waves. As always, the brain's electric and magnetic fields remain pre-eminent. By examining the factors conserving the electronic configuration, one can extrapolate as to the processes driving the absorption-impulse and inactivation-emission phases.

The electronic configuration's highly-symmetrical laws, forces and constants determine, or otherwise reflect, the movement of the electrons in their orbits, the structure of the resultant organic molecules, whether such molecules are **anabolized** or **catabolized**, and the absorption and emission spectra of the brain cell's myriad components. Theoretically, the primacy of the electric and magnetic fields is conserved at the sub-atomic, atomic, molecular, cellular, anatomical, celestial and astronomical scales. Therefore, **configurational theory** hypothesizes that radiotransmission dominates the human brain's internal workings, just as the electric and magnetic fields govern the innermost workings of the atom – and that those fields remain resolutely on top, irrespective of the multifarious organisms evolving beneath them. Making conspicuous life-forms a corporeal shadow, cast by conceptual forces.

Figure 30: configurational theory

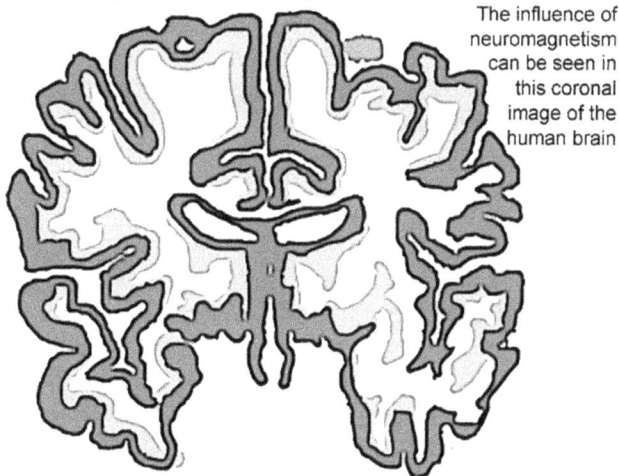

The influence of neuromagnetism can be seen in this coronal image of the human brain

Configurational theory argues that sub-atomic forces are evident at the macro-scale - not least of which, the primacy of electric and magnetic fields

Configurational theory sees neuromagnetism as influencing the brain's physical structure – an anatomical structure which relies, more than any other, upon absorption and emission spectra (see Fig. 30). Accordingly, one's **multivibrational matrix** represents the business end of so much radiotransmission, neurotransmission and arteriotransmission. What then happens, as regards the brain's radiofrequency exchanges, forms the basis of general and applied cognitive pragmatics. By mapping this elusive multivibrational matrix, one can effectively discard more antiquated physical representations of the brain, leaving only one's commanding 'ethereal spirit'. A 'spirit' which is the product of electric and magnetic fields working in concert. And so, whilst **corporealism** encourages material explanations of reality, amid an ever-growing list of particles, **conceptualism** charts these all-important electromagnetic energies.

- Gene attribution

Configurational theory argues that **mitosis** and **meiosis** conserve the atom's sub-atomic laws, forces and constants. Which implies that the electric and magnetic fields play a pivotal role in sexual reproduction and associated cell division, and therefore in **binary fission** and **gamete formation**. In other words, by conserving the electronic configuration – or, more accurately, the relative importance of all the factors providing for the same – we find determinism at the heart of every single biological process, including morphogeny, ontogeny and phylogeny. It's profoundly ironical, but conceptual reality's ability to fabricate asymmetries is endlessly conserved, by virtue of corporeal reality's unassailable symmetries. But, having curiously afforded hominins the freedom to stray, humans then meet with **configurational determinism**.

Life's common denominator, the cell, undergoes cell division as part of a preordained cycle. Which, in the case of eukaryotic cells, means growth and development, via mitosis (or else, gamete formation, via meiosis, for the express purpose of sexual reproduction). Such cell division is driven by the self-same fields which corrupt, for better or worse, gamete formation and binary fission. Thus, the conceptual shapes the corporeal through top-down electromotive and nucleomotive forces, rendering the corporeal ever more conceptualizing through time. Much of that reconfiguration of matter being driven by overlapping magnetic fields, which amalgamate in three-dimensions to produce variable electromotive and nucleomotive forces. Forces which then machinate at each and every level – not least, the molecular.

And, what the conceptual has engineered, with the assistance of genetic encryption, is remotely-accessible brain tissue. That landmark scientific triumph – whereby the human brain's latent telepathic potential was ingeniously unlocked, sometime between the years 1913 and 1947 (see **Destructive Interference, 2014**) – having happened without the safety-net of ameroliptical control, let alone the wholesale extinction of sapient tendencies.

Fortunately, when Nietzschean crises, cold-hearted nihilism and negative emotions vie with non-negotiable laws, forces astronomically greater than human agency promote positive feelings, sound insights and balanced responses. Of course, when space was plentiful, and resources abundant, *Homo sapiens* thrived. However, that **diegesis** has since been turned on its head, with the competitive exclusion of bias and logic already under way.

PART VI

Genome

Chapters 16-18

Ch16: **Dominant and recessive**

- **Laws of inheritance**

Gregor Mendel (1822-84), the Austrian monk who grappled with the laws of inheritance, was irrefutably aequipine. Not only did he isolate phenomena, employ mathematics and soundly interpret data, he also possessed an estimable disposition.[41] However, his true worth only surfaced after his death, thanks to the chance discovery of a published paper. In this paper, Mendel introduced his monohybrid inheritance experiments, in which he isolated contrasting traits in **pure-breeding** plants. In one such experiment, he crossed pure-breeding tall and dwarf plants, and found that the resulting seeds produced only tall plants. One of those first generation of tall plants was then self-pollinated, producing tall and dwarf plants in the ratio of 3:1. As one could combine and separate characteristics, it became clear that the mechanism involved must be discrete or particulate. Mendel had, to his posthumous credit, discovered **genes**.

Mendel's work established that plants and animals possess a **genotype**, comprising all the organism's genes, together with an expressed **phenotype**, or outward appearance. If we call the gene for tallness 'T' and the gene for dwarfism 't', then the genotype of Mendel's first generation, mentioned above, would be written as 'Tt' (with the resultant plants all possessing a tall phenotype). As that first generation possessed contrasting genes, they were all genetically **heterozygous**. When one of those heterozygous plants was self-pollinated, three-quarters of the second generation possessed the tall phenotype, due to having either a 'TT', 'Tt' or 'tT' genotype. As for the remaining quarter, they produced a dwarf variety, due to their genotype being 'tt' ('TT' and 'tt' being genetically **homozygous**). That second generation genotype was heterozygous and homozygous in the ratio 1:1, but the expressed phenotype was tall and dwarf in the ratio 3:1. Evidently, 'T' was the **dominant** trait, and 't' the **recessive**.

Significantly, that recessive gene could only express itself as a given phenotype in its homozygous form, a form called a **double recessive** (which, in the example quoted, would be written as 'tt'). The technical term for different versions of the same gene is **alleles**, and they're particularly interesting when examining **sex-linked** characteristics. The human **sex chromosomes** are labeled X and Y, with gamete formation in men producing spermatozoa with either an X or Y chromosome. Thus, the spermatozoon determines the sex of the child, boys being **heterogametic** (X-Y) and girls being **homogametic** (X-X). For argument's sake, let's imagine that the gene for balanced-mindedness only exists on the X chromosome and is recessive. Hypothetically, an ameroliptical nature would only ever be expressed in females, because only women would possess that all-important double-recessive.[42]

Figure 31: Sex chromosomes (filial ramifications)

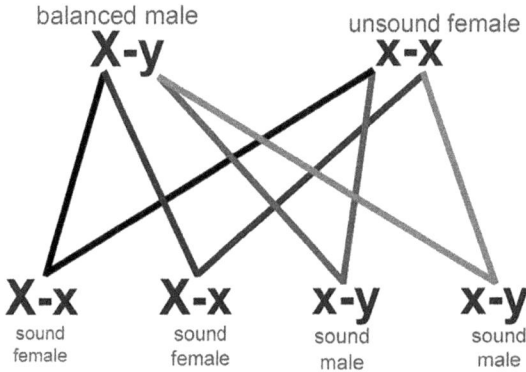

balanced male
unsound female

X-y   X-X

X-x   X-x   x-y   x-y

sound
female

sound
female

sound
male

sound
male

The filial ramifications, i.e. resultant offspring, should a pure-breeding aequipine male and pure-breeding sapient female have children, are speculated upon above. All the children appear perfectly sound, in spite of having an unstable mother.

Of course, we all know that the sexes are far more varied than this, due to wide-ranging differences in men's and women's CNE-axes. That said, we later give a plausible gametic reason why men are more disposed to sapient responses? For now, let's imagine that a woman whose CNE-axis makes her unsound is afforded the label 'x-x', and that a man whose CNE-axis makes him well-balanced is afforded the label 'X-y'. The **filial ramifications**, in the event of them having children together, are speculated upon in the accompanying illustration (see Fig. 31). As subsequently explained, it may be genetically impossible for women to replicate the most biased male responses. Nevertheless, pure-breeding *Homo sapiens* comprises 'x-x' and 'x-Y', whilst pure-breeding *Homo aequipondium* comprises 'X-X' and 'X-y' (with the former pairing-up for biased and logical reasons and the latter for reasons which are sound and balanced).

- Macro-molecular machinery

Macro-molecular machinery transcribes a gene from a small section of DNA, producing **messenger RNA (mRNA)**. This mRNA then exits the cell's nucleus, whereupon it affiliates with another piece of macro-molecular machinery, called a ribosome, which translates the strand's encrypted data. To recap, a gene is copied from one strand of the DNA double helix, in a process termed **transcription**, producing mRNA (then, the portion of that gene which codes for a specific characteristic, otherwise known as the **cistron**, gets copied by an organelle called a ribosome, in process known as **translation**, producing a protein). That relentless synergy, between nucleic acids and proteins, constituting configurational determinism's most significant achievement. Accordingly, macro-molecular proteins, with a **structure-function** custom-made to machinate under the influence of overlapping magnetic fields, help nucleic acids to produce yet more proteins.

The three-dimensional configuration of proteins, engineered by translation, strongly influences their biological roles (with those molecular assemblages frequently possessing areas of pronounced **positive** and **negative** charge, together with **hydrophilic** and **hydrophobic** properties). In practice, it's the actual size of the supporting carbon framework, relative to those all-important **functional groups**, which ultimately determines the protein's biological role. Those properties, or functional groups, enabling complex proteins to machinate under the influence of magnetic and electric fields, sufficient to transcribe the code in DNA, decipher the same by means of a ribosome or 'glue' amino acid chains together using peptide bonds. Thus, long before life had established itself, proteins had already achieved a conceptually-driven self-articulating purposeful structure, aided-and-abetted by nucleic acids.

A single gene codes for one whole protein, or, at the very least, a polypeptide chain. A typical protein contains twenty different amino acids, which, if not ingested, must be synthesized via anabolic pathways (some involving transamination). Translation borrows from **semiconservative replication**, whereby the DNA double helix 'unzips', and freely available bases, or nucleotides, attach themselves sequentially, producing a whole new DNA double helix. What differs, in addition to it taking place outside the cell's nucleus, is that one *isn't* dealing with a strand of DNA, but rather, a strand of mRNA – and that it's not nucleotides which affiliate, but rather, entire **transfer RNA (tRNA)** molecules, and their assorted amino acid appendages. It's those amino acid appendages which become bound together, producing a fully-functioning three-dimensional protein.

To all intent and purposes, mRNA and ribosomes behave like a **catalyst,** or **enzyme**, in-so-far as they bring together amino acids. However, this extraordinary form of **catalysis** requires instructions, which stipulate precisely what to catalyze. In the context of this book, such catalysis fabricates a cognitive-nervous-endocrine axis, allied to that all-important multivibrational matrix. Rewind the biological clock, and we arrive at a time in the primordial past when nucleic acids catalyzed polypeptide coats, in the manner of viruses. Successfully catalyze the forerunner of today's ribosomes, however, and the resultant macro-molecular machinery, together with the nucleic acids themselves, begin to synergistically 'coevolve'. Thus, evolution and extinction emerged at that point in time when proteins and nucleic acids attained an equal standing.

- ## Chromosomes, chromatids and centromeres

Only by conceptualizing **chromatids** can we make cell division comprehensible. Likewise, it's important to understand that every cell derives from another, and that their growth, function and development progresses according to a cell cycle. With those provisos, let's examine cell division in greater detail. In human conception, two **gametes**, produced by meiosis, fuse, producing a fertilized **ovum** (those original gametes possessing 23 chromosomes apiece). Consequently, the fertilized egg, or unicellular human, possesses 46 chromosomes. Each chromosome then shortens and thickens, revealing itself as 2 chromatids, joined at a **centromere** (like a letter 'X'). Thus, there are 92 chromatids, which pull apart, producing two cells, each containing 46 chromatids, whose genetic structure resembles the original 46 chromosomes. Hey presto, one now has two cells with identical genetic material, in a process termed mitosis.

Mitosis is concerned with preserving one's genetic code, whereas meiosis is all about mixing genes up, by means of **crossing over** and **independent assortment**, in order to create unique offspring. In the case of meiotic crossing over, two **homologous chromosomes** (lying side-by-side, in an 'XX' configuration) touch at certain junctures along their four chromatids, sufficient to exchange chunks of genetic material. How those two chromosomes then line-up, in order to separate, is arrived at randomly – in other words, they independently assort. By this means, dominant and recessive genes (or alleles) are distributed amongst the gametes produced by meiosis. However, that still leaves open the question of how those dominant and recessive genes, and even entire **sex-linked local groups**, are handled by the cell's macro-molecular machinery.

Some have suggested that dominant and recessive genes betray the ease with which fully **functioning proteins** are synthesized. But this is only one of several plausible explanations. Alternative propositions include **epistasis**, in which monitor genes switch alleles from a dominant state to a recessive state, and back again, according to the laws of natural selection. Additionally, dominant and recessive features may be an offshoot of **differentiation** (with cells discriminating not just on the basis of functionality, but also between genes of comparable function). Therefore, all cells ruthlessly discriminate, *vis-à-vis* cistron expression – with 'dominant' and 'recessive' being relative terms, related to the cell's anticipated function. That is to say, the cell's own proteins determine what is, or isn't, important.

As regards maturation and development, the cell cycle has three principal stages: 1) interphase (which includes growth, protein synthesis and regular functioning); 2) **mitotic cell division** (in which two fresh nuclei, containing indistinguishable genetic material, are produced); and 3) cytokinesis (during which the cytoplasm divides, separating the two nuclei, and thus completing the formation of two completely new cells).[43] Cell differentiation probably has its origins in interphase, with its emphasis on functionally-relevant protein synthesis. Contrary to media hype, active brain cells never multiply.[44] Thus, the neurons driving the mind cycle have no cell cycle, only regular functioning – such that if stages 2 and 3 were to occur, it would probably signal mental impairment. Consequently, configurational theory perceives electric and magnetic fields as driving forward regular cognitive function, aided-and-abetted by an absence of cell division.

Curiously, science has convinced itself that the escalating scale of things somehow subverts sub-atomic forces, and that those phenomena which govern the internal workings of the atom don't have the same relevance once we get beyond cells, tissues, organs and people. This causes some to dismiss out-of-hand the proposition that magnetic and electric fields play a pivotal role in biology, let alone gravitational and dispersional forces. Somewhat controversially, this book postulates that those sub-atomic forces are deterministically conserved, most conspicuously within galactical media. That symmetry serving to explain such things as conscious self-awareness, and even aspects of mitotic and meiotic cell division. Indeed, it's that very symmetry which makes galaxies fraught with **intelligent life** (with some anthropoid life-forms pro-actively learning and suitably adapting quicker than others).

* Trinomial nomenclature

Binomial nomenclature is an indispensable Linnaean device, but a device which struggles in the face of hominin reasoning (be it epipoliotical, hermeneutical, epivevatical or ameroliptical). Consequently, a Linnaean paradigm shift is looming, as the cognitive, nervous and endocrinal activities of *Homo sapiens epipolioticus* and *Homo sapiens hermeneutics* vie with those of *Homo aequipondium epivevatics*. This conceptual shift is driven by the hyponymic demands of an inescapable adaptive radiation in human **neurophysiology**, coupled with a pressing need to anticipate the impact of the same on earth's biosphere. Accordingly, this hominin rivalry looks set to be managed from above by a global governance complex's neuro-cognitive wing. As one might expect, not all aequipines are pure-breeding, presenting balanced-minded individuals with an interesting sociological challenge.

Arguably, there are four male genotypes, expressed as four neurotypes (1. balanced, X–y; 2. sound, x–y; 3. logical, X–Y; and 4. biased, x–Y). But *only* three female genotypes, expressed as three neurotypes (1. ameroliptical, X–X; 2. epivevatical, X–x; and 3. hermeneutical, x–x). Tellingly, there is no biased female genotype, only a socially-constructed approximation. On a point of pedantry, we should actually say that there's no epipoliotical female genotype. *Ergo*, it's impossible for the female genotype to produce a Las Vegas style mass-shooter (such as the one responsible for the Route 91 Harvest Festival massacre, 2017), save-and-except for protracted radicalization, or the reprehensible use of radiofrequency effects (see Fig. 32). Regrettably, *Homo sapiens epipolioticus* will always prize aggression, stockpile weapons and cruelly outrage, due to a low-attention span, limited intelligence and unmediated basal responses.

Imagine, for argument's sake, an isolated population of humans, consisting solely of biased males (x–Y) and balanced females (X–X). The first filial generation would comprise sound females and logical males. Later, the second filial generation would abruptly diversify, with biased and logical males pairing-up with sound and balanced females. Biased males then achieve sexual primacy,

Figure 32: Sex-linked local groups (male and female genotyopes)

genotype	MEN	WOMEN
ameroliptical (balanced)	X-y	X-X
epivevatical (sound)	x-y	X-x
hermeneutical (logical)	X-Y	x-x
epipoliotical (biased)	x-Y	

Arguably, there are four male genotypes (expressed as four distinct neurotypes), but only three female genotypes (expressed as three possible neurotypes). Accordingly, there appear to be genetic limits, as regards women's biological capacity to commit unconscionable acts of extreme violence.

through the ruthless appropriation of economic, political, military and ideological power, with sound females their only mates. By the third filial generation the sociopolitical superstructure is consciously pairing logical females with biased males, in order to advance the regime's prestige agenda. Punishingly, by the fourth filial generation all newborns are unsound – and will remain so *ad infinitum*, because that priceless 'X' chromosome has vanished from the gene pool.

What the above scenario powerfully illustrates is the ability of the sociopolitical superstructure to fundamentally alter the underlying genetic substructure. But, by the same token, ill-judged structuration, peer-to-peer communication and misguided liberalism can also undo the good work done by nature and/or 'domestic repression', leaving society at the mercy of neurotypic deficiencies. As luck would have it, cruciform calculus progressively selects those individuals who, in the phraseology of **trinomial nomenclature**, we'd label *Homo aequipondium amerolipticus*. *Homo aequipondium amerolipticus* being a pure-breeding human, which can only ever produce balanced-minded offspring, all other neurotypes having vanished from the gene pool (effectively turning the above example on its head). Moreover, with those unbalanced **non-autosomal** 'x' and 'Y' chromosomes competitively excluded, the cruciform structure then serves to inhibit their recurrence.

•   Demes, clines and complexes

Demes are genetically isolated sub-populations (characterized by stabilizing fecundity or selection) which, in the event of **environmental resistance**, may become **clines** (subject to progressive fecundity or selection). Beyond varieties of human, we encounter **races**, or ethnic groups. And, beyond them, we meet with sub-species and **species**. Varieties are distinguishable on the basis of minor **phenotypic** or **neurotypic** differences. Races and sub-species are identifiable as a consequence of phenotypic, neurotypic and/or genotypic factors. And, species are distinguishable on the basis of substantive genotypic transmutations. Presently, races of human are distinguishable solely on the basis of their phenotypes, with the scientific community having no commonly agreed genetic basis for racially classifying people.[45] Indeed, it's difficult to see why one would want to categorize people on the basis of race, other than to address pernicious racial bias.

Classifying individuals on the basis of their reasoning would be more profitable, beginning with those who unsoundly perceive genotypic racial differences. Only by abandoning our discomfiture in respect of neurotypic variation, axial radiation and cognitive sub-speciation – and by anticipating the shocking ecological ramifications of hopeless inactivity – can we begin to craft the overarching structures needed to sustain constructive human advance. The previous book (**Destructive Interference, 2014**) concluded that "*as the culture we keep is so often facile, particularly amongst the most sexually-active, coevolution seems doomed to be a relatively superficial process*". Given that pessimistic assessment, this follow-up book's cruciform strategy comes with a commanding clinical model of the ameroliptical mindset, its non-negotiable cruciform laws growing society on the basis of balanced-minded principles.

According to the Hardy-Weinberg Principle, allele frequencies – in a sizable deme, subject to random mating – will remain constant, provided there's no selection, mutation or gene-flow (notwithstanding that selection can be stabilizing).[46] Expressed more bluntly, if one or more of those conditions is flouted, adaptive radiation, sub-speciation and wholesale evolution will follow. As those conditions ought to be flouted, why not guide the biological substructure in the direction of balanced-mindedness. Accordingly, whilst *Homo aequipondium amerolipticus* constitutes a prospective deme, other categories of hominin rank as prospective clines. An anthropological cline being a genetic sub-population which competitively adapts, due to the rewarding ramifications of progressive fecundity, profitable mutations, beneficial gene-flow and perspicacious forms of mating.

- Coevolution's beating heart

Semiotics is coevolution's 'beating heart'. Quite simply, it does more than anything else, as regards driving that priceless sociological cycle (with humanity's ecological niche conceptualized, customized and crafted through a medley of symbolic socializing processes). Which both explains and justifies analytic philosophy's penetrating linguistic analysis, science's repudiation of specious conjecture and feminism's mounting optimism in respect of informed disaggregated exchanges. The best of women's instincts, like the feelings uniting the most erudite epistemic community or philosophy's soundest logicians, closely mirroring the fact that our collective future is won or lost through reason. Therefore, critically evaluating society's upcoming utilization of language is vital. Our future being less about artillery, air power and atomic bombs, and more about cogitation, categorization and competitiveness.

In the absence of remote manipulation, mental telepathy and the superimposition of persons, and discounting both **psychokinesis** and **telekinesis**, the CN-cycle draws exclusively upon afferent and efferent signals travelling to and from the individual's own central nervous system. Depending on one's multivibrational matrix, a person may be minded to afford afferent signals more weight and importance than efferent signals, making them empiricist by nature. Rationalists, on the other hand, perceive their mind cycles as outputting authoritative interpretations, causing them to place greater weight and importance on efferent signals. Those with a rationalist mindset experience an 'intoxicating' chemical rush, due to the manner in which their CN-cycles action their NE-cycles. Conversely, empiricists experience delayed endocrinal responses, due to a deplorable lack of anticipation – their CN-cycles keeping their NE-cycles in the dark until the last decisive moment.

Thus, the less cognizant action their NE-cycle empirically, and in ways which are emblematic of the classic **'fight or flight' response**. Given the deceptive nature of reality, undue empiricism leads to unsound induction and unintelligible inaction (as happens when legislators fail to grasp that social learning is corroding due process, or when members of the public fail to realize that their attention is being disingenuously deflected). *Homo aequipondium*, through the expedient of hypothetico-deductive scrutiny, subjects its own and others' analytical thought processes to negative feedback, thereby tempering the rationalist's weakness for mania. Some might facetiously claim that *Homo sapiens* possesses an '*RE-cycle*', rather than a CN-cycle, its radiological and electrochemical exchanges favouring theocratic explanations. What's clear, is that the female CNE-axis provides for a constructive coevolutionary dynamic – one in which women are *dominant*, rather than *recessive*.

Ch17: **Women's best instincts**

- Proscriptive reasoning

Aequipine forms of reasoning are **prescriptive**, i.e. they're types of analysis which ought to be encouraged, while sapient forms of reasoning are **proscriptive**, i.e. they're types of analysis which are best avoided. In an ideal world, the human genome would support only prescriptive forms of reasoning, making aequiponimity a biological certainty. In broad terms, sound analysis is precise, unambiguous and thorough, while unsound analysis is imprecise, ambiguous and lacking. Moreover, while the former presents uncertainty in terms of arithmetical probabilities, the latter dons an arrogant, doctrinaire or hubristic attitude. In terms of actual behaviour, unsound minds *'favour punishing; using reward selfishly, and without obvious limits'*. The very antithesis of what selfless intra-family concern has caused the vast majority of women to become over the course of geological time.

Proscriptive reasoning can be classified as either epipoliotical or hermeneutical, depending on just how lacking the given argument, statement or rhetoric happens to be. **Epipoliotical reasoning**, for example, betrays poor judgement, due to a marked absence of methodology, deliberation or reflection (which, in extremis, merges with the incoherent ramblings of the criminally insane). Such individuals frequently possess poor ancillary skills, such as those associated with observation, analysis and decision-making. Not to mention, an excess of *'non-sequiturs'* (meaning that their conclusions rarely follow logically from their premises). Additionally, the epipoliotical may over-compensate with emotive language, preferring swearing, obloquy and invective, in lieu of cogent self-expression. Or, they may retreat into ideological assumptions and the telling use of ascription.

Hermeneutical reasoning, whilst more structured, is no less reliant upon unsound premises. Such arguments contain, amongst other things, false correlations, invalid comparisons and unwarranted leaps, whilst otherwise maintaining a high degree of internal consistency. What's more, the logically-minded habitually **connote**, leading to imprecision, ambiguity and doubt. Also, by arguing by association, employing latent messages and engaging in implicit arguments the hermeneutical are born propagandists. Indeed, they may avoid expert opinion, primary source material and objective sampling in order to make a convincing case – a case built upon unsound consensus, rather than informed disagreement. To that end, one also encounters inexcusable tautology, sleight of hand and deflective language (their hubris being chemically amplified by a cocktail of motivational highs, many associated with the dubious chemistry of group bonding).

- Prescriptive reasoning

Prescriptive reasoning can be classified as either epivevatical or ameroliptical, depending on the degree to which the given argument, statement or proposition appears sound or balanced. **Epivevatical reasoning**, for example, corresponds with scientific deliberation, and the crafting of sound arguments by those possessing superlative ancillary skills (that is, faculties associated with observation, analysis and decision-making). Accordingly, such reasoning contains all the necessary propositions (e.g. variables afforded the notations 'p', 'q', etc), together with sufficient logical constants (e.g. connectives such as 'then', 'not', 'and', 'or', 'if and only If', etc), to reliably convey factual information. Additionally, sound minds value first-rate source material, respected authorities and, in the case of scientific research, replicability, reproducibility and repeatability.

Ameroliptical reasoning, whilst no less sound, appears less reductionist. Or, to be perfectly precise, it's more competent at objectively appraising the context within which knowledge is soundly amassed. Although possessing much the same cognitive abilities as the epivevatical mindset, it augments the same with balanced-mindedness. The ameroliptical mindset is therefore a priceless transferable skill, whether one works in the field of early learning, peer review or conflict resolution. Within the law, for example, it can raise doubts regarding the actions of the police or judiciary, with a view to safeguarding **due process**. Within science, it can critically-appraise the sociological or humanitarian impact of strategic concealment. And, in the field of child development, it can promote the optimal conditions for personal growth.

An ameroliptical person is ever mindful of the political climate, vested interests and human nature. Consequently, well-balanced observers understand implicitly that sapiency subverts the truth (possibly leading to neuro-cognitive effects being applied in ways which run contrary to the interests of minors). Quite unlike the blinkered clinician, hopelessly marooned within a reductionist mindset, the balanced-minded protagonist might actually borrow from proscriptive reasoning, in order to flag-up otherwise intractable problems (for example, when a well-meaning polemicist, stymied by the mitre paradox, triggers a timely debate). An independent observer is then faced with an **attributional dilemma**, i.e. whether to attribute that person's *prima facie* inductive pronouncements to their disposition or to the epistemological demands of their given situation?

A woman's best instincts would say *"employ hermeneutics, if only to place the issue on the agenda, and in the absence of sounder alternatives"*. Occasionally, therefore, a challenging attributional debate surrounds a person's use of tautology, sleight of hand and deflective language. That is to say, if they're used conscientiously, in a manner suggestive of situational attribution, their utilization may be excusable. However, if they're used impulsively, or in a partisan way – possibly to the detriment of civil liberties – then the person undoubtedly possesses defective reasoning. Just as an aequipine male is compelled, on occasions, to use violence, a balanced-minded female might, on occasions, be forced by official disinformation to use proscriptive practices – especially if telepathically-induced effects have been willfully misappropriated, and the truth is being callously suppressed.

• Evolution of coevolution

The orthodoxy is surprisingly close to the truth when it holds that women are well-placed to craft a balanced environment for children and men well-equipped to respond to external threats in an ill-balanced manner. Whilst structurally very complex, both male and female brains can be understood in comparatively simple terms (see Fig. 33). The CN-cycle, or mind cycle, pertains to the most recent part of the **forebrain** (that is, the majority of the **telencephalon**). Conversely, the NE-cycle comprises a small portion of the telencephalon, called the basal ganglia, together with the more antiquated part of the forebrain (that is, the diencephalon), together with the **midbrain** (**mesencephalon**), **hindbrain** (comprising the **metencephalon** and **myelencephalon**), autonomic nervous system and endocrine glands.

The question arises, how does the CN-cycle attach itself to the NE-cycle? The simplest analogy one can give is that of the press-stud, whereby the limbic system (positioned beneath the **cerebrum**, within the telencephalon) accommodates the basal ganglia and diencephalon, etc, as they are, figuratively-speaking, 'pushed upwards'. Whereupon, the CN-cycle's limbic system becomes intimately attached to the NE-cycle's basal ganglia.

Imagine, for a moment, the seat of analytical thought, memory, emotions, motor learning and conscious movement, clicking neatly into the seat of autonomic and hormonal communication. As you might imagine, there's an important body positioned at the focus of this limbic and basal confluence, termed the thalamus, which relays sensory information to various parts of the brain, where it can be routinely processed (all assuming, of course, that one's senses aren't being cruelly deprived).

Figure 33: Mind, brain and body

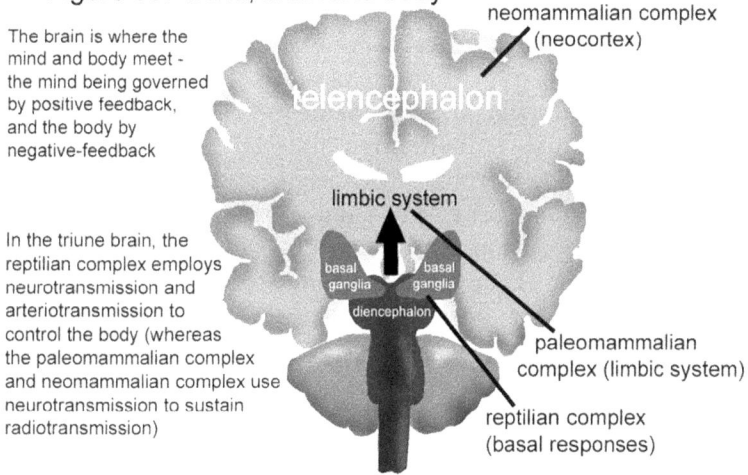

The brain is where the mind and body meet - the mind being governed by positive feedback, and the body by negative-feedback

In the triune brain, the reptilian complex employs neurotransmission and arteriotransmission to control the body (whereas the paleomammalian complex and neomammalian complex use neurotransmission to sustain radiotransmission)

neomammalian complex (neocortex)

telencephalon

limbic system

basal ganglia

basal ganglia

diencephalon

paleomammalian complex (limbic system)

reptilian complex (basal responses)

The limbic system, in addition to generating and regulating personality and emotions, also assists in the formation of memories. The basal ganglia, for its part, generates and regulates behaviour, aiding the acquisition of **unconscious competency** in a broad range of tasks. Together, the basal ganglia and diencephalon (or, more specifically, the thalamus and hypothalamus) constitute the NE-cycle's control centre or 'brain'. Above them, functionally-speaking, we meet with positive-feedback, in the form of the divergent CN-cycle. Significantly, the neurotransmitters which arouse this higher cognitive function (for example, dopamine, serotonin, oxytocin and endorphins) are identical to the ones produced and secreted by the NE-cycle's diencephalon, basal ganglia and brainstem. In other words, the mammalian brain exploits chemical messengers with decidedly ancient origins.

Therefore, the entire telencephalon, with the exception of the basal ganglia, behaves in a computational mammalian manner. In fact, one proposes reclassifying the basal ganglia, so that it forms part of the diencephalon. This anatomical reclassification reflects the fact that the NE-cycle behaves like a pre-mammalian central nervous system, replete with its very own 'brain' or 'expressed will', including chemically-charged basal responses. As for the newly-redefined telencephalon, that exists to lay-down nubits of memory, logic-gated conditioning and idiosyncratic ways of reasoning. Moreover, it does so with the support of regular **amino acid transmitters**, such as glutamate, aspartate, glycine and gamma-amino butyric acid (which, together, inhibit or excite contiguous neurons). Of course, when thinking about disposition one must account for personality and behaviour, whilst ever conscious of the corrupting impact of circumstance.

Although bodies such as the limbic system's amygdala are associated with rage, actual rage is a product of chemicals produced and secreted by the nervous-endocrine cycle. Making the amygdala a computational initiator of that quick-tempered response, rather than the physiological response itself. It just so happens that the telencephalon of *Homo aequipondium amerolipticus* processes sensory information in a well-balanced manner, only rarely triggering the body's rage response. And, even then, only once the facts have been established. Accordingly, the CN-cycle, or mind cycle, informs personality, whereas the NE-cycle, or **body cycle**, informs behaviour. Clearly, a painstaking surgical analysis of the brain's limbic system and basal ganglia is called-for, in order to categorize their subtlest anatomical details.

In conclusion, the body cycle evolved as a system in its own right, whereas its counterpoint, the mind cycle, represents a mammalian upgrade (providing for both Pavlovian and operant conditioning, the forerunners of today's complex reasoning). In effect, you socially-construct emotions using your **telencephalonic mind**, which your body then interprets using a cornucopia of neurotransmitters and hormones (chemical messengers with the propensity to reward or dissuade, via elation and alarm). Prescriptive and proscriptive forms of reasoning reflect upon that telencephalonic upgrade, but remain hostage to the body's antediluvian chemical heritage (such that an epivevatical person may respond in an ill-balanced manner). As for one's **brain**, that spans both the radiological mind and the hormonal body – making it an indispensable electrochemical hub, custom-built to mediate between otherwise incongruent cycles.

- *De novo* synthesis

When a nucleotide is erroneously substituted during semiconservative replication, that alteration may have no immediate effect, but nonetheless increase the chances of subsequent mutations, or it may cause a completely different amino acid to be incorporated into an ensuing protein. If that earlier genetic blunder involved a regulatory sequence of **deoxyribonucleic acid**, then its impact on protein production could be even greater. Moreover, if the whole of translation were to shift, causing each triplet of bases to be read in an offset manner, a much larger mutational effect would occur. What appears to have happened, is that non-autosomal local groups (governing a range of radiological, electrochemical and hormonal functions) have mutated into specific male and female axes. Begging the question, have those sex-linked genetic mutations been favoured by natural forces or by human agency?

Forces far exceeding human agency favour plasticity, and that plasticity has the appearance of human agency being favoured. However, in the case of *Hominini* things became ecologically confused, due to the destructive impact of personality, with its heavy emphasis on self-indulgent social conditioning – personality, more than any other factor, serving to disrupt otherwise stable ecosystems. However, there remains, colloquially-speaking, 'an animal' within us all – replete with mindless instincts, autonomic responses and **steroidal hormones**. Those instincts, responses and hormones impinge upon the seat of personality, thereby augmenting its more negative aspects. *Homo sapiens'* end is nigh, because, quite frankly, it's not enough of an animal to be moulded behaviourally, nor is its personality sufficiently aequipine to coevolve successfully. In truth, it's an intermedial species, facing a bleak future.

To render *Homo sapiens* replacement **clinically unsound** requires, at the very least, **mental disorder**, **heavy metal toxicity**, congenital defects and/or the malign application of telepathically-induced effects. That is to say, the range of factors needed to alter, for example, an epivevatical mindset to a hermeneutical one are considerable. In the case of mental disorder, such as dementia, one would require a substantial breakdown of the multivibrational matrix. For its part, heavy metal toxicity (involving lead, mercury, manganese, cadmium, etc) is likely to produce a cascade of **hormonal imbalance**, not-to-mention permanent neurological impairment. Additionally, certain inherited conditions disrupt **cholesterol** homeostasis, thereby compromising every one of the brain and body's communication systems. Therefore, we must familiarize ourselves with those factors which impair balanced minds – not least of which, heavy metal contamination of the food-chain.

Cholesterol homeostasis is really quite interesting, with every component of one's CNE-axis requiring this biologically-essential **sterol** (a molecule carried in one's bloodstream by lipoproteins). Unfortunately, the brain is incapable of absorbing cholesterol directly from the blood stream, due to the so-called 'blood-brain-barrier', therefore it must synthesize cholesterol *in situ* using far simpler molecules, in a process termed *de novo* **synthesis**. Unsurprisingly, cholesterol metabolism, regulation and excretion dominate clinical research. Basically, unless cholesterol levels are optimal, memory, learning and behaviour will all suffer (as will **neurogenesis, synaptogenesis** and **reproductive health**). That's because, in spite of high blood cholesterol causing heart disease and arteriosclerosis, the cognitive, nervous and endocrine systems would all cease functioning without it.

Neurosteroids, produced from cholesterol, have an excitatory or inhibitory effect on the mind cycle. And so, irrespective of whether your reasoning is proscriptive or prescriptive, cholesterol metabolism supports the very core of your being. In fact, the more we learn about cholesterol homeostasis, particularly in relation to that explosion of neonate neurological development termed **exuberant synaptogenesis**, the more medicine suspects that **neurodegenerative disease** may derive from its antithesis. However, getting cholesterol levels correct is just the start, the brain and body must then use cholesterol to synthesize the 'master' steroid hormone **pregnenolone**. Pregnenolone is then used to produce glucorticoids, mineralcorticoids and the sex hormones (namely, progesterone, testosterone and oestrogen), substances which antagonistically regulate, amongst other things, mineral balance, electrolyte concentrations, pregnancy, etc.

Irritability, fatigue, anxiety, depression and mood disorders all derive from this metabolic balancing act being 'out-of-sync'. In extremis, one might even meet with an existentialist crisis, arising from the wholesale corruption of one's CNE-axis. What Sigmund Freud (1856-1939) termed the **id**, i.e. the human brain's primordial source of unconscious drives and desires, is, in point of fact, the pre-mammalian 'brain', or reptilian complex (comprising the basal ganglia, diencephalon, etc).[47] That id possesses a will of its very own, it having been a self-regulating 'animal' in its own right. The question, is whether the telencephalonic mind was ever meant to be anything other than the id's servant, computing behavioural changes in response to environmental resistance. The answer, is that only by becoming pure-breeding aequipines will humans warrant global pre-eminence – a case of **superego** winning-out over **ego**.

• Additional learning needs

The European Commission's president might argue that the European Union is a socioeconomic environment first, a political system second and a military power third. Contrast that with the United Kingdom, which presents itself as a major military power, stationed within a highly politicized Europe, for whom society is a burden to be reduced. Today, the UK appears smitten with military prowess (having become, somewhat paradoxically, a place where one gets reprimanded for upholding the 'responsibility to protect' principle). By promoting federalism, Brussels favours a rewarding socioeconomic environment over the self-interested deployment of military power – a position which greatly appeals to those occupying the political middle-ground, and therefore to those building a more prosperous and stable Europe.

It would be a grave mistake to allow a biased and logical dissolution of the European Union, without some means of subordinating the chief protagonists. Great Britain, by framing itself as prototypically 'non-European', risks being governed less from the political Centre, and more from the socialist Left and capitalist Right, by turns. An anachronistic scenario which risks plaguing British politics with unsound types, fearful of exploitation, insurgency and post-imperial trends. Predictably, the EU's stance finds favour with those sensitive to other's needs, while the UK's self-interrogation regarding its own wants finds favour with those sensitive to their own desires. Which would explain why 80% of UK women, aged 18-24, voted to remain in the recent **EU referendum** (a position which contrasts sharply with those for whom competing has become an unshakable defect).

Observing children at play enables educationalists to assess their motor, social, intellectual, emotional and interaction skills, together with their grasp of non-verbal communication. According to these criteria, peer-to-peer communication places you at the top of the heap, intellectually, psychologically and emotionally. And yet, that social learning often serves to undermine sound and balanced thinking, leading to conclusions at odds with the facts. Therefore, is it really stretching credibility to view unsound interaction as a condition on the **autistic spectrum**? That is to say, do popularist agendas highlight **additional learning needs**? Brussel's **pluralist 'tabula rasa'**, or blank democratic slate, neatly provides for a constructive coevolution – but can't succeed if the body politic is largely sub-aequipine. Indeed, as with every other social structure, the European Union must somehow address the corrosive impact of *Homo sapiens.*

• Bridging heaven and earth

Cruciform calculus, like its differential and integral namesakes, is mathematically pragmatic and numerically abstracted. Admittedly, the term 'logicism' can be used pejoratively to connote an unsound use of algebra. However, provided one's approach is arithmetically sound, employing only such pragmatism and abstraction as is necessary, then an objective analysis can be arrived at. Without question, the limits of cruciform coevolution are delineated by structurating agents and socializing structures, beyond which one meets with a mandatory humility arising from the ramifications of configurational determinism. Accordingly, while aequiponimity involves the fabrication of social structures with sufficient plasticity to be shaped by natural laws, sapiency comprises an insistent structuration, driven by insatiable subjective yearnings.

Like it or not, genes and memes are tested by unforgiving trial and error, with *Homo aequipondium* possessing the most prescriptive memes and fittest genes, and *Homo sapiens* possessing the most proscriptive memes and weakest genes. Post-radical feminists, who concede that their axial advantage stems from conditions transcending mere liberation, are well-placed to craft the optimal conditions for growth – not least, by acknowledging the role played by socially-constructed emotions. Therefore, a women's less lateralized processing of sensory and extrasensory information serves as an ideal starting point when meeting the four measures of soundness, appraising the context within which hypothetico-deductive research takes place or crafting a well-balanced social environment for children.

Looked at from the perspective of connubial dualism, the superimposition of comparable neurotypes parallels structuration and socialization.[48] For example, in its heteronnubial form (as opposed to its **homonnubial** form) a balanced-minded female might serve as the '*socializing structure*', with her male companion adopting the role of a '*structurating agent*'. Indeed, with their multivibrational matrices amalgamated, they might, in the manner of this book, elect to publish their combined thoughts. Question: given the intimacies involved, which is more important, the ability to relax or the ability to excite?[49] Exciting or not, heteronnubial dualists might jointly determine the facts, oversee a scarcity of information, critically evaluate one another's thinking and objectively interpret the available evidence. Clearly, to avoid peer-to-peer failings, only aequipines should be wedded in this manner – after all, anything less than aequiponimity would invite **criminal solicitation** by voyeuristic sapients.

Ch18: A *'godly'* conclusion

• Overwhelming symmetry

The dark sciences argue against a 'big bang', much preferring an interpretation in which stellar nucleosynthesis and **black holes** are counterbalanced by deep-space nucleosynthesis and dark matter (the universe's mutually attracting protons and mutually repelling electrons generating both strong and weak nuclear forces).[50] At the focus of this astrophysical symmetry, one finds the electronic configuration – an electron's movements being governed by the electronic constant, electromotive force and electrostatic attraction (and a proton's movements, and hence the orientation of nuclei, being governed by the nucleonic constant, nucleomotive force and electrostatic attraction). Given that our galaxy's central supermassive black hole retains the energy emanating from its diminishing core, it might, as one's heteronnubial co-author contends, have begun to relax its gravitational grip in respect of its distinctive spiralling arms.

The nucleomotive force exists because the nucleonic constant gives atomic nuclei no alternative but to re-orientate themselves under the influence of fluctuating magnetic fields. The electromotive force exists because the electronic constant gives electrons no alternative but to move under the influence of fluctuating electric fields. This affords those energetic fields the conceptual upper-hand, with corporeal matter transmuting according to their will. Neuromagnetism amends the electronic configuration's shells, subshells and orbitals, sufficient to generate person-specific **electron jumps**; enabling the human brain to sustain a frequency modulated multivibrational matrix, and with it an *'imperfectly inscribable'* telencephalonic mind. A mind which has successfully mastered the manipulation of electricity, both as an immediate source of energy and as a binary input within computational systems built around Boolean operators.

Forget classical physics, electrostatic attraction is inversely-proportional to energy, not distance. That is to say, the greatest conceivable symmetry comes from ambient energy levels being so low that electrostatic attraction increases astronomically. That electrostatic attraction producing, in the first instance, atoms – but, in extremis, a form of entropy dominated by neutrons. For now, galactical energy levels permit electrons to flow like a current, as they dissociate electrostatically from their parent atoms.

Conventional current flow, as it appears on an illustration of an electronic circuit, runs from the positive to negative terminal. However, **real electron flow** (that is, the path taken by the electrons themselves) runs from the negative to positive terminal. Configurational theory rejects the contention that these real and conventional currents generate magnetic fields, as that would imply that particles have primacy over fields.

According to configurational theory, a variable electric field produces a potential difference. Which, to the untutored eye, looks indistinguishable from a current generating a magnetic field (indeed, the field effect transistor uses an electric field to control an appliance's electrical behaviour). Similarly, a fluctuating magnetic field exerts a force on a current-carrying wire's atomic nuclei, whilst simultaneously influencing the electric fields surrounding the same. Configurational theory further argues that electrical currents are harnessable due to them being a corporeal manifestation of ambient energy levels and associated electrostatics. That said, the 'one true' competitive exclusion would much prefer that electricity was exploited by sound and balanced minds, rather than biased and logical ones – with even the technologies themselves being classifiable as 'aequipine' or 'sapient', owing to their biased algorithms, logical algebra, sound computations and balanced outputs.

- Digital determinism

First conceived of in the 1970s, America's space shuttle had, by the beginning of the 21st century, become the short-lived backbone of the International Space Station (a technological mainstay which, prior to Elon Musk's (1971–) enterprising SpaceX venture, was charged with ferrying personnel and supplies to that permanently-manned zenith of human ingenuity). Beginning in the mid-1990s, 15 nations, including the USA and Russia, began piecing together that unprecedented platform in space. A classic example of technological innovation and non-partisan scientific collaboration discontinuously driving forward scientific discovery. Contemporaneously, personal computers, mobile phones and the internet firmly established themselves across all seven continents. At the heart of this latter-day renaissance was the omnipresent semiconductor, whose discovery provided for logic-gated electrical devices.[51]

The most commonplace semiconductor, the transistor, was introduced in 1947 (just as America began concealing telepathically-induced effects from the world, and as the Soviets began interfering with surface traffic between the German Federal Republic and Allied-controlled sector of Berlin). The transistor was neither a perfect conductor, nor a perfect insulator, its electrical conductivity being controlled by electrical currents, arising under the influence of electromagnetic fields. Thus, the Cold War (1946–89), **digital determinism** (1947–) and the tactical concealment of neuro-cognitive effects (1947–) were all contemporaneous with the rolling-out of these transistor-based technologies. Those transistors having transformed discrete circuits, integrated circuits and microprocessors into functionally-complete electronic systems, whose very logic made them available to unsound factions – hence the political tensions emerging at that time.

A microprocessor takes **binary inputs** (in the form of high and low voltages, corresponding to ones and zeros) and channels them through millions of transistors printed onto a 'silicon chip's' surface, in order to output an answer in binary notation. That is, Boolean operators (acting in the manner of so many logic gates, latches and switches) process the information fed to them in the form of currents. Accordingly, in the same way that propositional calculus evaluates propositions using **logical constants** (e.g.→,~, &, ∨, ↔) and mathematics evaluates numbers using **logical operators** (e.g. +, -, ÷), microprocessors evaluate binary inputs using **logical connectives** (e.g. AND, OR, XOR, NOT, NAND). However, whilst analytic philosophy endeavors to integrate semiotics, mathematics and computer logic, the results could still be 'sapient'. That is to say, *Homo sapiens* is perfectly proficient at coding, can network digitally and often texts whilst driving.

Bistable multivibrators have two stable states, enabling that handy piece of electronic circuitry to be switched from a 'zero' to a 'one', and *vice versa*. In that way, a computer can memorize huge amounts of binary data. Humans and computers both utilize **constants**, **operators** and **connectives**, in order to evaluate propositions, numbers and currents, sufficient to calculate, analyze and respond. However, both must be deemed 'sapient', until proven sound. The answer, therefore, isn't for humans to emulate robots, but rather, for software engineers, like the rest of us, to eschew the merely logical. Seen from the perspective of *Homo aequipondium ameroliptic us*, even epivevatics isn't enough. That having shunned sub-aequiponimity, they're then forced to question the cold-blooded pursuit of knowledge, often for its own sake, by scientists whose application of constants, operators and connectives falls short of being properly well-balanced.

• Frege, Wittgenstein and Russell

German analytic philosopher Gottlob Frege (1848-1925), having convinced himself that the system of argument and proof at the heart of mathematics was intellectually unsurpassed, proceeded to devise predicate calculus. Later work, by Ludwig Wittgenstein (1889-1951) and Bertrand Russell (1872-1970), established a comparable branch of linguistic logic, termed logical atomism. Frege, Wittgenstein and Russell were all driven by a common desire to deconstruct people's statements – in much the same way that Andreas Vesalius (1514–64) had sought to dissect the human body. This philosophical and linguistic revival amounted to actively promoting verifiable facts as the '*atomic*' building blocks of sound '*molecular*' arguments, all sensibly '*bonded*' together by sentential operators (the Vienna Circle later consolidating their ideas under the umbrella term **logical positivism**).

The Vienna Circle exemplified the wisdom of *Homo aequipondium epivevatics*, with its hardline insistence on **verificationism**. Aesthetics, ethics and metaphysics were all roundly condemned for being too subjective, unduly conjectural or egregiously rationalist. What's clear, is that the hypothetico-deductive cycle would've met with their approval, due to its strict scientific methodology. However, the Vienna Circle's fundamental weakness was its lack of **attributional awareness**, unsound pragmatics being attributable to either disposition or circumstance. For example, when an otherwise balanced person converses unintelligibly, due to aphasia arising from an ischemic stroke, circumstance *is* everything, making their statements anything but meaningless. Likewise, the mitre paradox arises because logical positivism counts for nothing if the facts regarding remotely-induced trauma cannot be formally established.

Had Frege, Wittgenstein, and Russell met with undisclosed neuro-cognitive violence, logical positivism would've sealed their fates (leaving balanced-minded protagonists to sow the seeds of an attributional dilemma, through their polemical use of proscribed formulas). The problem is that technologically-advanced states can manipulate a citizenry's entire corpus of knowledge, given their top-down command of purportedly factual information, leaving the likes of the Vienna Circle struggling to verify or disprove official accounts. That's why epivevatical individuals possess only partial cognitive pragmatic competence. It's not that anyone questions their command of hypothetico-deductive reasoning, *per se* – it's just that Anne Sullivan (1866-1936) and Helen Keller (1880-1968) would've been more discriminating, as regards establishing the facts, evaluating assertions, addressing incomprehension and attributing wisely.

The Vienna Circle had faith in the 'atomization' of language, believing that a tried and tested linguistic formula would somehow generate a range of compelling philosophical solutions. However, an argument isn't won or lost on the basis of the argument itself – such that those who dismiss the truth, or accept falsehoods, on the basis of poor or convincing arguments are technically unsound. Therefore, 'atomizing' statements makes you little more than logical, without truly exhaustive verificationism. Given those intractable problems, balanced-mindedness feels an awful lot like sensory deprivation. But, if not logical positivism, then what? As luck would have it, mother nature has gifted us an ostensibly child-centred solution, one which acknowledges the damage wrought by *laissez faire* liberation – and which seeks to reduce, as far as possible, the 'mist and fog' arising from unsound actions and behaviours.

•   Ameroliptical perspective

Even when *Homo aequipondium amerolipticus* profitably bridges *'heaven and earth'* with sound algebra – sufficient to **predict** the future or **retrodict** the past – they'll still require a pedagogic device transcending mere fact-finding, in lieu of radiating ethics. As we've seen, the migration of analytic philosophy towards mathematical logic and computer science, by way of logical positivism, leaves society burdened with a broader philosophical vacuum. Cruciform calculus, through its promotion of balanced-mindedness, does much to fill that void, but a person's commitment to the same is likely to be a reflection of their sociological perspective. For example, structuralists, like many **positivists**, view society as a product of its permanent structures, economic systems and data acquisition. Conversely, social actionists, like many **anti-positivists**, view society as emerging from the unstructured murk of subjective bias, logical deliberation and social learning.

Positivists see sociology as a science, which accords with the view that predictive modelling is aided by sound and balanced structures. Anti-positivists, conversely, see sociology as unscientific, which tallies with the notion that predictive modelling is confounded by grassroots bias and logic. Nevertheless, positivists and anti-positivists both agree that any form of personal expression which goes beyond hardline verificationism is, by definition, irrational, subjective and inductive. Where they differ, however, is that positivists and structuralists view all of this as subsidiary, provided that sound and balanced structures prevail, whereas anti-positivists and social actionists see the social environment as inevitably frustrating scientific inquiry. That is to say, short-term **predictions** and **retrodictions** can still be made in respect of unsound elements, it's just that sub-aequiponimity thwarts long-term scientific analysis.

Phenomenologists are extreme anti-positivists, who approach sociology from an existentialist, some would say sapient, direction. *Ergo*, anti-positivism and phenomenology are microsociological perspectives, which see no evidence of *Homo sapiens*' competitive exclusion, and thus dismiss out-of-hand the notion of scientific sociological modelling. Opposing them, is this book's extreme positivism, termed the **ameroliptical perspective**, which is a sociological position founded upon the macrosociological premise of increasing aequiponimity, amid diminishing sapiency, sufficient to make scientific modelling easier through time. Both **phenomenology** and **amerolipticism** see the human mind as key – what makes them irreconcilable, however, is that ameroliptical individuals see cruciform calculus as coevolving grounds for scientific optimism, whereas phenomenologists see society as fundamentally resistant to scientific expression.[52]

Mammalian plasticity paved the way for Pavlovian and operant conditioning, such conditioning giving rise to a branch of psychology termed **behaviourism** (behaviourism's preoccupation with conditioned behaviours affording human psychology the superficial appearance of a science). However, attempts at understanding *Homo sapiens*, using behaviourism, met with only limited success – a sub-aequipine's willful empiricism either imperfectly mimicking the expectations of behaviourists or its chronic rationalism owing more to emergent **cognitive psychology**. Above and beyond *Homo sapiens*' inborn deficiencies, **anti-behaviourism** reflects the fact that behavioural responses will prove deceptive, if a person has been subject to sustained radiological abuse.[53] *Homo aequipondium amerolipticus*, however, sweeps away all of these uncertainties – giving rise to an altogether more predictable branch of cognitive psychology, based upon sound and balanced thinking.

- Truth tables

Avowals (propositions), **voltages** (binary inputs) and **values** (numbers), linked by constants, operators and connectives, comprise the inputs from which the past and future are extrapolated. Tellingly, the discrete electronic circuit exemplifies the problems facing those struggling to predict and retrodict. Imagine, for argument's sake, that one has a transistor functioning as an **AND gate**, such that its output 'Q' is only a high voltage if both the 'A' and 'B' inputs are high voltages (see Fig. 34). The hobbyist can *only* retrodict 'the past' inputs if the present output is high and an 'AND event' is known to have occurred. And, they can only predict 'the future' output, if they know that an 'AND event' is going to occur and have full knowledge of the present high inputs. Proving that time, as one's heteronnubial co-author poetically asserts, is an impenetrable mist, blinding our respective senses and causing many a momentary belief to be positively-reinforced.

Figure 34: Avowals, voltages and values

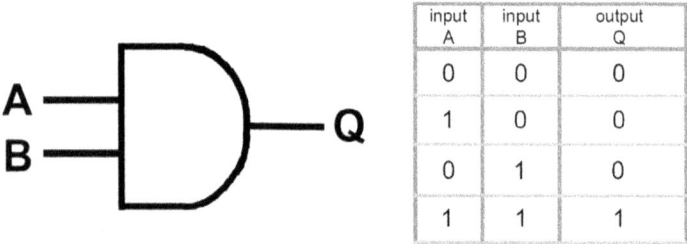

input A	input B	output Q
0	0	0
1	0	0
0	1	0
1	1	1

Propositions, inputs and numbers are the means by which the past and future are extrapolated. In the absence of the scientific method and contemporary source material one must know, for example, that an 'AND' event has occurred, or that such an event is about to occur (and have full knowledge of the inputs and/or outputs)

The soundest way to retrodict, is to use every pertinent output to eliminate all but the most conceivable inputs, which one then infers must have happened. Thus, computed retrodictions, regarding past events, are comparable to reverse engineering an entire electronic circuit, together with all its inputs, using only the observed outputs (bearing in mind that one would also have to deduce every single event separating the present outputs from the past inputs). Even if that were possible, it would probably be computationally draining. Nevertheless, cruciform calculus starts the process off by determining the most pertinent outputs, relative to what is being retrodicted. With prediction, of course, those outputs become the inputs. Clearly, prediction and retrodiction, by definition, denote something far more cutting-edge than hypothetico-deductive reasoning and evidence-based history, however elusive that aspiration.

Basal responses are arguably more predictable than socially-constructed emotions – whether it be a love of mathematics, despair at ascription, anger at injustice or a fear of inequality. That is to say, extrapolating in respect of past, present and future neurotypic responses necessarily involves accounting for emotions. Achieve that, and *Homo aequipondium amerolipticus*, armed with unprecedented powers of prediction and retrodiction, would tower over epivevatical science, the soundest of historians and anyone struggling with contemporary source material. One thing is for certain, as the past, present and future all conform to configurational theory, there's little doubting their conserved characteristics. Indeed, one could do worse than worship those aspects of conceptual and corporeal reality which *are* conserved – the ramifications of which appear very positive indeed.

Gazing into the future, the repudiation of faith is all very well – but seldom, if ever, does its secular replacement adequately stress forgiveness, or even **due process**; with the news and social media routinely amplifying, or otherwise augmenting, questionable memories, unsound extrajudicial practices and scarcely concealed *sub judice* (a position, it must be said, which is at odds with the predictive and retrodictive rigour outlined above). One could argue that if you're aequipine, religion does you no harm, and that if you're sapient, secularism does you no good. Secular or sacred, we need to stop marketing unrepresentative models of ourselves and the cosmos, and begin urgently revising both in order to surmount the most predictable of calamities. One's co-author hits the right note, when she asserts that amerolipticism relieves society of the need to predict and retrodict, with *Homo sapiens* no longer exacerbating the deficiencies inherent within mathematics.

- Guaranteed free speech

The closest that the United Kingdom has come to a constitutional guarantee of free speech, at the time of writing, is its ratification of the **European Convention on Human Rights** (Article 10, of which, states ... "*everyone has the right of free expression. This right shall include the freedom to hold opinions and to receive and impart information and ideas without interference by public authority and* *regardless of frontiers"*). The writer, ever conscious of the human rights ramifications of telepathically-induced effects, took it upon themselves to commence lobbying on these issues whilst working as an **Officer of the Supreme Court**, within the jurisdiction of England and Wales.[54] Beginning in the late 1990s, **IMPART** (and later **WOMEN IMPART**, a successor campaign inspired by one's heteronnubial co-author) began inviting an urgent re-examination of the rationale of those applying radiofrequency effects.[55]

Much of that urgent re-examination stemmed from the fact that we're living in an interstitial age, with its casual subversion of textbook protocols. All being well, this book's promotion of neocentric stratification will inspire a constructive 'race-to-the-top' (a 'race' which is fought on the basis of avowals, voltages and values, all tellingly 'glued' together by constants, operators and connectives, and in a manner which is symptomatic of syntax, semantics and pragmatics). Such are the ravages of pecuniary and prestige reasoning – especially where it results in the corruption of cognitive, nervous and endocrinal processes – that the imposition of non-negotiable cruciform laws looks like a '*god send*'. Those laws reflecting upon the alarming nature of contemporary technical know-how, and the pressing need to grow the human genome before it's too late.

And so, whilst men stockpile the accoutrements needed in the event of a prospective 'race-to-the-bottom', women can intelligently capitalize upon their sex-linked axial advantages, by honing-in on what really matters – namely, the radiological, the electrochemical and the hormonal. The resultant cogitation, categorization and competitiveness serving to redefine the true nature of power. Such that mainstream economic, political, military and ideological powers become embroiled in a progressive phase transition; one in which top-down electromotive and nucleomotive forces readily decompose the same, amid the promise of absolute prosperity. Significantly, this book, like its well-intentioned precursor, is the product of two convergent minds. Such that if women do take over the world – becoming, to all intent and purposes, an '*almighty influence*' – it will have begun with one's heteronnubial co-author's commanding ameroliptical presence.

Appendix

Glossary

All listed in alphabetical order (ignoring the prefix 'the' and any preceding numbers). The term **dark sciences** denotes the scientific models developed in this book.

A	Definition / description
Abelard's dictum (philosophy/law)	"*By doubting we come to inquiry; through inquiring we come to perceive the truth*". See also, **Cogito, ergo sum**.
Absolute advantage (politics/economics)	**Aequipine** individuals possess an absolute advantage, arrived at through **sound deduction** (**sapient** individuals having, at best, only a **comparative advantage**). See also, **aequipine economics**.
Absolute poverty (economics/sociology)	Absolute poverty (e.g. food-bank reliance, malnutrition and indebted penury) isn't remediable through **market** forces, only through socialism's **decomposition of capital**. See also, **relative poverty**.
Absolute prosperity (economics/sociology)	Capitalism, tempered by **socialism** (and the **decomposition of power**), produces absolute prosperity, and with it conditions transcending pure economic well-being. See also, **relative prosperity**.
Absolute refractory period (neurology)	The brief rest period during which **neurons** can't be re-actioned (hypothesized, in this book, as forming part of the **inactivation-emission** phase, when **thoughts** and **brain waves** are emitted).
Absorption (dark sciences)	Overlapping **magnetic fields** strengthen an atom's **ground-state** magnetism, initiating the absorption of a **photon** (a case of incident radiation generating the conditions for its own absorption).
Absorption-impulse Absorption-impulse phase (dark sciences)	The absorption-impulse phase is **endergonic**, requiring **energy** from **thoughts**, **brain waves** and/or **neurotransmitters**. By catalyzing enzymatic reaction within a neuron's **cytoplasm** (with supplementary energy from **ATP**), this phase generates **nerve impulses**. See also, **inactivation-emission**, **unstable multivibrator** and **CN-cycle**.
Absorption profile/s (dark sciences)	The person-specific pattern of **absorption** and **emission** (arising from an individual's **neuromagnetism**), which makes **telepathically-induced effects** possible.
Abstraction (economics/sciences)	The **economic** equivalent of **pragmatism**, in which details are ignored, in order to focus exclusively on the problem. More broadly, abstraction occurs when **complex adaptive systems** are expressed in discrete terms (for example, when they are visualized).
Acculturation (sociology)	Pertains to those instances when immigrants appropriate a host culture. See also, **transculturation** and **socialization**.
Acetylcholinesterase (biochemistry)	An **enzyme** which deactivates a **nerve cell**, in the manner of **negative-feedback**; such that life-threatening **positive-feedback** will arise (e.g. spasms, convulsions and respiratory problems) if nerve agents suppress its actions.
Active transport (cellular biology)	Active transport aids sub-cellular **homeostasis**, or **negative-feedback**, and is dependent upon **ATP** (such that cyanide, which disrupts ATP synthesis, prevents a cell from maintaining its physiological **set-point** against a concentration gradient). See also, **NE-cycle**.
Actor-observer bias (attribution theory)	Arises when negative **behaviour** in others is attributed to their **disposition**, but similar behaviour, arising in oneself, would be attributed to the demands of the situation. A political elite, prone to actor-observer bias, will dismiss hard-pressed families as having flawed dispositions. Right-wingers often failing to grasp that capitalism is fraught with difficulties, which privileged advocates of the markets ought to sympathize with, rather than merely denigrate.

Acute nonlinear event (mathematics)	Mathematicians unjustifiably promote chaos, calamity, catastrophe and criticality, as **life**, if anything, discriminates against the same by harnessing Gibbs free **energy** in pursuit of stable communities. Accordingly, the term acute nonlinear event better describes a sudden shift in the fortunes of a **complex adaptive system** (the ramifications of which may be constructive *or* destructive).
AD 1913-47 (science/history)	The timeframe within which the human brain's latent telepathic potential is thought to have been realized (this **neuro-cognitive** potential having been tapped into without the safety-net of **balanced-minded** control or the **extinction** of **sapient** tendencies). See also, **Destructive Interference, 2014**.
Adaptively-radiate Adaptive radiation	The propensity of all species to mutate, sufficient to fill every available **ecological niche** (with natural selection discriminating against the prevailing **phenotypes** or against unusual alternatives).
Additional learning needs (sociology)	Educators commonly evaluate motor, social, intellectual, emotional and interaction skills, together with a child's grasp of non-verbal communication, in order to discern additional learning needs (and yet, it is those very skills which enable those without learning disability to engage in corrosive forms of **social learning**, i.e. what this book colloquially terms **dumbing-down**).
Adenosine triphosphate (ATP)	ATP's production is **endergonic**, i.e. it cannot be synthesized spontaneously in nature (as with naturally-occurring **compounds**, such as water), but must be metabolized within the organism, using available energy. It's 'second' only to **nucleic acids** and **proteins** in making **life** possible, in-so-far as it enables life-sustaining physiological **set-points** to be established and maintained (ATP being the plant or animal cell's immediate source of **energy**).
Adjectival social stratification Adjectival stratification	Adjectival social stratification denotes the twenty-four configurations which are possible when society is stratified according to the adjectives **biased**, **logical**, **sound** and **balanced**. The term **adjectival stratification** denotes a hegemonic democracy which is rich in social action, giving it a latent potential for **neocentric stratification** or **pathological stratification**.
Admissible (law)	Only legally-acceptable evidence is deemed admissible in law. One might argue that evidence, as well as allegations, ought to be refutable, affording the law parity with **science**. After all, the law could find itself advancing towards some as-yet-unproven nonsense on the back of weak **evidence** (with much time and money saved if it were to confine itself to refutable evidence and allegations). See also, **mitre paradox** and **inadmissible**.
Aequipine Aequipine economics Aequiponimity	Applied to those who are **sound** and **well-balanced**, such aequiponimity affording those with superlative **critical thinking skills** an **absolute advantage** (not just economically, but more broadly). Aequiponimity, particularly **balanced-mindedness**, appears to be a function of **sex** in women and **gender** in men (that is, **biologically-produced** and **socially-constructed** respectively). See also, **sapiency**.
Affectation (semantics)	'Affect' is defined as "to act upon or influence, especially in an adverse manner". 'Affectation' (that is, 'affect' + suffix 'ation') should be defined as a deviation from natural or innate behaviour, for reasons which are more diverse than merely seeking to impress, e.g. "the affectation was due to an external agent acting in an adverse manner".

187

Afferent signals (neurology)	Nerve impulses which are conveyed **electrochemically** towards the brain. See also, **empiricist**.
Agent/s	Individuals possessing a given **neurotype**.
Aggregate (power)	When faced with anarchy and disorder, a pluralist aggregation of **power** is called-for (which, in the case of postwar Germany, took the form of consolidated **economic** power). See also, **concentrate**, **defragment** and *pluralist 'tabula rasa'*.
Aggregate demand (economics)	A macroeconomic device which pools the demand for all goods and services into one convenient figure.
Alleles (genetics)	Variant forms of a given **gene**.
Almighty influence *'almighty influence'* (philosophy)	**Competitiveness** isn't decided by humans, it's a **conceptual** judgement arising beyond ourselves, and which ultimately determines which **species** of human inhabits our global niche. In effect, **energy** equals **mass** times the **speed of light** squared, in-so-far as 'E' and 'mc²' are interchangeable via **deep-space nucleosynthesis** and **stellar nucleosynthesis**, but the manner in which they mutually influence is anything but equal, with energy having the upper-hand. Thus, energy creates a humanity-shaping conceptual reality, one in which an individual may be personally-shaped by **connubial dualism** (making one's inamorata an *'almighty influence'*). See also, **'one true' competitive exclusion**, **tensor-poiesis** and **E=mc²**.
Amelioration	Occurs when a word or phrase assumes a more positive meaning.
America	See **United States of America**.
Americano-Sunni bloc / Americano-Sunni bloc subordination	The Arab Gulf states and Turkey comprise a bloc of predominantly **Sunni** Muslim countries, all of which are allied to **The West**, in-spite of being plagued by anti-Western factions, e.g. Islamic State and al-Qaeda. Perhaps it's Sunni insecurity, regarding Western military support, which causes them to invest so heavily in armaments and munitions. Any or all of the states comprising this Americano-Sunni bloc may meet with **subordination**. See also, **Russo-Shia bloc**.
Ameroliptical	See **ameroliptical reasoning**.
Ameroliptical perspective	See **amerolipticism**.
Ameroliptical reasoning	Ameroliptical reasoning is less reductionist than scientific reasoning – or, in truth, more competent at appraising the context within which such knowledge is soundly amassed, sufficient to apply that knowledge in a **well-balanced** manner. See also, **ameroliptics**, **amerolipticism**, **prescriptive reasoning** and **epivevatical reasoning**.
Amerolipticism Amerolipticist/s	Amerolipticists are extreme **positivists**, who, by virtue of **cruciform laws**, predict increasing **aequiponimity** and diminishing **sapiency**, sufficient to model society with increasing **macrosociological** proficiency. Amerolipticism is therefore the antithesis of **phenomenology**, with the evolving human **mind** providing for a compelling **science** of society. See also, **ameroliptics**, **ameroliptical reasoning** and **anti-positivists**.
Ameroliptics (scientific field)	The scientific study of **balanced-mindedness**, as regards its origin, impact and expression (from *amerolipsia*, Greek for impartiality). It doesn't follow that a person conversant with **coding**, or computer programming, will necessarily comprehend **epistemology**, any more than it follows that **scientists**, confident of their facts, will articulate themselves ameroliptically. See also, **amerolipticism.**

Amino acid/s Amino acid transmitters	Amino acids are the molecular sub-units from which **proteins** are constructed (all possessing an amino group and a carboxyl group, but differing in their functionally significant 'R' groups). In addition to making proteins, they also serve as **neurotransmitters**.
Amplitude (physics)	The maximum displacement of a wave from what would otherwise be a flat calm or rest position.
Anabolic Anabolized	Anabolic processes are **endergonic** (energy-demanding biological reactions, which build tissues and organs using energy from **ATP**).
Analytic **philosophers** Analytic philosophy	Philosophers who promote **formalized logic**, and who examine statements, propositions and arguments looking for logical inconsistencies. One could argue that they are more **empirical** than **rationalists**, preferring scrutiny over metaphysical contemplation.
AND gate (Boolean operator)	A **logic gate** (where the output is a high voltage *only* if both inputs are high voltages).
Antagonistic **hormones**	**Hormones** which have the opposite physiological effect, enabling the body to self-regulate around a **set-point**. **Cholesterol** is used to synthesize the 'master' steroid hormone **pregnenolone**. Pregnenolone is then used to produce glucorticoids, mineralcorticoids and the sex hormones (substances which antagonistically regulate a panoply of functions). See also, **homeostasis.**
Anthropobionomics Anthropocene	Anthropobionomics is the study of the genus *Homo* in relation to its environment; with the anthropocene being the present geological age, within which *Homo sapiens*' activities serve to affect the nature of the rock strata subsequently laid down (whilst its replacement, *Homo aequipondium*, achieves global primacy). Arguably, 'plasticene' might have better described an anthropocene littered with plastic!
Anthropogenic **factors**	The impact of the genus *Homo* upon the natural environment. See also, **anthropobionomics** and **anthropocene**.
Anti-austerity (politics/economics)	The **Right-wing**, with its penchant for **dispositional attribution**, tends towards **austerity**, i.e. **monetary policies** which risk **unemployment** and **absolute poverty**. The **Left-wing**, with its proclivity for **situational attribution**, tends towards anti-austerity, i.e. **fiscal policies** which risk **inflation** and higher taxes.
Anti-behaviourism (psychology)	Anti-behaviourists accept that **cognitive, nervous** and **endocrinal** responses will be misleading, if a person has been subject to malign radiological interference or intolerable stress. See also, **behaviourism** and **attributional dilemma**.
Anti-capitalist Anti-capitalism	An ideological position predicated upon the notion that **market** logic is **unsound**. However, whilst **capitalism** is technically unsound, it remains an unavoidable precursor to **socialist** interventions.
Anti-exploitation **bias** (sociology)	A grassroots bias, predicated upon the belief that **state** oppression, social inequality and political subjugation are pernicious and widespread. See also, **social action crucible**.
Anti-humanism Anti-humanist/s	Anti-humanists see **humanists** as culturally **biased**, in-so-far as faith in human virtue appears to stem not from people, but from the **social structures** imposed by a select few. Other critics of humanism cite its nebulous principles and lack of adequate introspection. For a more rigorous approach to social advance, see **neurotypic sub-classifications, ameroliptics** and **cruciform calculus.**

189

Anti-insurgency bias (sociology)	A ruling class bias, motivated by the belief that any electorate denied commanding strategic intelligence, and which is prone to perilous levels of **cognitive bias**, cannot be guaranteed to enter into self-disciplined lobbying. See also, **counter-insurgency**.
Anti-positivist/s (sociology)	Anti-positivists see sociology as unscientific, with predictive modelling confounded by endemic **bias** and **logic**. See also, **positivists**, **phenomenologists** and **ameroliptical perspective**.
Anti-realism (maths/philosophy)	In mathematics, anti-realism views algebra as **rationalist**, rather than **empirical** (that is to say, **subjective** rather than **objective**).
Anti-Semitic policies Anti-Semitism	Policies which discriminate against Jewish people. Arguably, the Third **Reich** sought to liquidate **Zionism** *en masse* in order to win favour with the Arabs, sufficient to access Middle-Eastern oil. Answerable to Hitler, at that time, were **logical minds**, capable of deducing the strategic 'benefits' of the fuhrer's **anti-Semitic policies**, including the Holocaust itself. Suggesting that the minorities are particularly vulnerable to calculating **political logic**.
Apoptosis (neuroscience)	Apoptosis, or programmed cell death, renders roughly half of all human **brain** cells defunct. See also, **neurogenerative disease**.
a posteriori	**Empiricists** infer *a posteriori* from what they see (giving rise to **unsound induction**). See also, '*a priori' coniunctum 'a posteriori*'.
Applied cognitive pragmatics	That branch of **cognitive pragmatics** which focuses exclusively upon professional deliberation.
a priori	**Rationalists** infer *a priori* from what they cogitate (giving rise to **unsound induction**). See also, '*a priori' coniunctum 'a posteriori*'.
'*a priori' coniunctum 'a posteriori*'	The scientific '(a) + (b) method' (that is to say, if (a), which is falsifiable, cannot be empirically disproved by (b), then (a) stands). The right kind of '(a)' being refutable, and therefore indicating the most pertinent forms of '(b)'. See also, **hypothetico-deductive cycle, sound deduction** and **epivevatics**.
Arab-Israeli conflict (history/politics)	Following **Israel**'s establishment, in 1948, displaced Palestinian Arabs drew sympathy and support from neighbouring Arab states, creating a continuous state of political and military tension in the Middle East. See also, **UN Security Council Resolution 242**.
Arab League Arab nationalists	Founded in 1945 by Arab nationalists, the Arab league boasts a population and area comparable to the **European Union**. However, being a loosely-affiliated **military** bloc (ridden with **political** and **economic** differences, for whom secular society is an intolerable burden), it cannot be thought of as emulating European advance.
Archaea (taxonomy)	See **prokaryotic cells**.
Arteriotransmission	Transported via the blood stream, e.g. **hormones**. See also, **neurotransmission** and **radiotransmission**.
Artificial intelligence (computer science)	Computerized **intelligence** may be viewed as **sound** or **unsound, rationalist** or **empiricist, objective** or **subjective**. Given these distinctions, all artificial intelligence must be presumed **logical** (and hence **'sapient', until proven sound**). Additionally, one must avoid obscuring the growing capacity for **sub-aequipine** human error.

Ascription Ascriptive cleansing Ascriptive practices	The practice of ascribing individuals attributes at birth (placing them in, for example, a lower caste, position of domestic servitude, or else classifying them as socially inferior). In extremis, they become prey to politically-motivated ethnic and social cleansing (for the record, **neurotype** isn't ascribed, it's soundly adduced).
Asexual reproduction	Reproduction involving fission, budding and vegetative reproduction. See also, **sexual reproduction**.
Atomic mass (physics/chemistry)	An element's relative atomic **mass** is a weighted average of all its naturally-occurring isotopes, i.e. the **atoms** of that **element** which contain different numbers of **neutrons**.
Atomic model, dark cycle (dark sciences)	The **dark cycle atomic model** explains **atoms** using four particles (**photons, electrons, protons** and **neutrons**) and four fields (**magnetic, electric, gravitational** and **dispersional**). According to the **fourth law of thermodynamics**, electrons possess an **electronic constant** and **nucleons** a **nucleonic constant** (such that, an **electromotive force** acts on the electrons and a **nucleomotive force** on the nuclei, due to the actions of the electric and magnetic fields respectively, giving rise to **bonding, cleaving, absorption** and **emission**). Additionally, the nucleons mutually-attract and the electrons mutually-repel, owing to gravitational and dispersional forces. See also, **dark cycle theory of everything**.
Atomic weight	See **atomic mass**.
Atom/s	Atoms comprise three particles (**protons, neutrons** and **electrons**). As the number of protons and electrons is equal, the atom carries no electrical charge (losing or gaining an electron produces an ion).
ATP/ATP hydrolysis	See **adenosine triphosphate**.
Attribution Attributional awareness (critical thinking)	Attribution theory is concerned with how people attribute causes to their own and other people's **behaviour**, with the most astute possessing exceptional attributional awareness (that is, they soundly adduce whether **dispositional attribution** or **situational attribution** is applicable). See also, **attributional dilemma**.
Attributional dilemma/s (critical thinking)	**Pecuniary extremism, prestige extremism** and **social learning** commonly obscure systemic abuse. Such abuse may force **well-balanced** individuals to borrow from **proscriptive** practices, in order to raise awareness (creating an attributional dilemma, as regards ill-informed third-parties). See also, **attribution** and **prescriptive**.
Aufbau principle (chemistry)	The principle that **electrons** fill the **orbitals** of an atom sequentially (starting with the lowest **energy** orbitals, closest to the **nucleus**, first).
Austerity (economics)	Right-wing **monetarists** logically reduce the availability of money, by means of austerity, in order to suppress **demand**, and hence control **inflation** (demand often exceeding **supply**, creating an inflationary pressure). See also, **anti-austerity** and **wage-price cycle**.
Authoritative power	Characterized by subordinates who obey unquestioningly.
Autistic condition Autistic spectrum	Characterized by abnormal social interaction, inability to form friendships and repetitive forms of **behaviour**; autism is a stigmatizing label which belies the fact that many without the condition nonetheless possess deficiencies as regards **critical thinking skills** (deficiencies which are amplified by social interactions, destructive friendships and impulsive behaviours). See also, **additional learning needs**.

Autoimmunity	A misdirection of one's immune system, resulting in the destruction of one's own body (there are over eighty autoimmune diseases).
Automatic destabilizer (economics)	In its revised form, the **global financial system** would naturally increase **Left-wing** and **Right-wing** sensitivity to supply and demand perturbations. What's more, **NEURON** would further amplify that sensitivity, as regards subordinate economies which are guilty of aggravating climate change, exhausting natural resources and accelerating habitat loss. See also, **automatic stabilizer**.
Automatic stabilizer (economics)	An inevitable consequence of balanced **economic** interventions, an automatic stabilizer would reduce the **Centreground**'s sensitivity to **supply** and **demand** perturbations (with **NEURON** helping to sustain that rewarding equilibrium). See also, **automatic destabilizer**.
Autonomic Autonomic nervous system	That part of the peripheral nervous system which is subdivided into **sympathetic** and **parasympathetic** branches. These two divisions, which have their main centres in the hypothalamus, act in concert in order to control blood pressure, body temperature, heart-rate, etc.
Autonomous morality	Morality deriving from one's own character, and which is supplemented discontinuously by experience. Such morality is unlikely to diminish (unlike heteronomous morality, which is a conditioned reflex).
Autosomal (genetics)	In addition to the usual **sex chromosomes**, or allosomes, human cells typically possess twenty-two pairs of autosomes.
Avowals	Propositions.
Axial advantage (neurotypic standing)	When a person or group possesses superior hormonal, electrochemical and radiological responses, relative to another person or group. **Post-radical feminists**, who concede that their axial advantage stems from conditions transcending mere liberation, are well-placed to craft the optimal conditions for growth. That axial advantage reflecting a **supraordinate**'s real time processing of sensory and **extrasensory** information – **aequipine**s being well-served by their **radiological**, **electrochemical** and **hormonal** reactions, and **sapient**s being fatally weakened by the same. See also, **cognitive-nervous-endocrine axis**..
Axon Axonal input (neurology)	The axon conveys nerve impulses away from the nerve cell's body towards contiguous **neurons** (with dendrites receiving signals from other neurons). In a **unstable multivibrator** an adjacent axon serves as an input – that is to say, it inputs to a **nubit** of memory. See also, **dendrital output**.
Azimuthal 'spdf' quantum number model	The azimuthal quantum number is one of four variables which, taken together, describe an electron's quantum state (a state which determines whether an orbital is an s, p, d or f shaped orbital, etc).
B	
Bacteria	See **prokaryotic cells**.
Balanced Balanced mind/s Balanced-minded Balanced-mindedness	The scientific study of balanced-mindedness is termed **ameroliptics** (from *amerolipsia*, Greek for impartiality). **Ameroliptical reasoning**, or the balanced mind, objectively appraises the context within which knowledge is amassed, whilst soundly amassing such knowledge – making it naturally judicious and highly resistant to **fundamental attribution errors**. See also, *Homo aequipondium amerolipticus* and **well-balanced**.

Basal ganglia Basal responses	The basal ganglia (positioned beneath the **cerebrum**, within the presently defined **telencephalon**) aids movement, including the acquisition of **unconscious competency** in a broad range of physical tasks. Students of **behaviour** should commence with the basal ganglia, basal responses being **biologically-produced** reactions, driven by **NE-cycle** (and therefore subject to **negative-feedback**).
Behaviour	Behaviour reflects biologically-produced **basal responses** (being mediated by **personality**, which draws upon socially-constructed **emotions**). However, to avoid **fundamental attribution errors**, one mustn't ignore the impact of circumstance. See also, **cognitive-nervous-endocrine axis**, **disposition** and **basal ganglia**.
Behaviourists Behaviourism (psychology)	Behaviourism became popular in the 20th century, due to the influence of Russian physiologist Ivan Petrovich Pavlov (1849-1936) and American psychologist John Watson (1878-1958). Behaviourists are **empirical**, believing that incoming **afferent signals** direct behaviour (unlike **cognitive psychologists**, whose **rationalist** perspectives view behaviour as symptomatic of outgoing **efferent signals**, emerging from subjective **cogitation**). See also, **anti-behaviourism**.
Beta decay (dark sciences)	**Neutrons**, produced by **deep-space nucleosynthesis**, are vulnerable to beta decay (a form of radioactive decay, initiated by ambient energy, which produces **protons** and **electrons**). In the absence of ambient **energy**, **electrostatic attraction** increases astronomically; that electrostatic attraction producing, in the first instance, **atoms** (but, in extremis, a form of **entropy** dominated by neutrons).
Biased Biasing Biased-minded Biased mind/s	The scientific study of biased thinking is termed **epipoliotics** (from *epipolaiotis*, Greek for cursoriness). **Epipoliotical reasoning**, or the biased mind, is the least empirically disciplined and most crudely rationalist mindset of them all, giving rise to subjective behaviours (which, as **cognitive psychologists** will attest, contradict empirically-verifiable truths). See also, *Homo sapiens epipolioticus*.
'Big bang' (astrophysics)	The notion of a 'big bang' began with Georges Lemaitre (1894–1966), when he suggested that the universe's **energy** and **mass** were once concentrated in some primordial 'cosmic egg', prior to its explosive decay (our universe being an estimated 13.7 billion years old). The **dark cycle theory of everything** demotes the 'big bang' theory to a **hypothesis**, arguing that astrophysics may be edging towards an as-yet-unproven nonsense. See also, **dispersional fields**.
Binary fission (biology)	A term which some apply to the division of **prokaryotic cells**, but others use to denote **mitosis** in **eukaryotic cells**.
Binary inputs Binary notation (computer science)	A binary input consist of a series of ones and zeros, each one of which is termed a bit. Together, eight of these bits form a byte (by eight), which can represent a total of 256 values, simply by rearranging the positions of the ones and zeros. See also, **Boolean operator**.
Binomial nomenclature (taxonomy)	The acme of **hyponymic hierarchies**, which serves to differentiate between all living organism using only their two-part Latinate names (comprising the **genus** and **species**). Whilst binomial nomenclature relies upon differences in **phenotype**, this book introduces **trinomial nomenclature** in order to distinguish between members of the genus *Homo* in terms of their cognitive processing of information, i.e. hominins cannot be categorized by phenotype alone. See also, **neurotypic sub-classifications**.

Biochemical homology	Compares and contrasts the chemical reactions taking place within the cells of different **species**. By this means taxonomists have been able to fine-tune our broader classification of living organisms.
Bioelectrogenesis (evolutionary biology)	From electroneutrality, the cell evolved a range of electrical potentials, each one involving a polarized outer membrane. That harnessing of **positive-feedback** heralding the start of a **mind cycle**.
Biogeochemical Biogeochemical norms (ecology)	What is normal? Well, if you're a living organism it's a physiological **set-point**, which your **NE-cycle** strives to maintain. Borrowing from **Gaia theory**, life also helps to create biogeochemical norms, at scales far exceeding the organisms themselves. See also, **homeostasis**.
Biologically produced (genetic substructure)	In the context of **cruciform calculus**, the term biologically-produced denotes the **human genome**, and the **phenotypic** and **neurotypic** differences which emerge from the same. See also, **socially-constructed** (sociopolitical superstructure).
Biological singularity	Life on earth appears to have arisen only once, at a specific location on its surface. **Configurational theory** says that a commanding **atomic model** should enable us to unravel that genesis. See also, **constructive theory of life's origins**.
Biomagnetism	The magnetism associated with living organisms is infinitely variable due to: 1) the cell's synthesis of large organic molecules; 2) their diverse orientation and three-dimensional structure; and 3) the impact of their overlapping magnetic field lines. See also, **neuromagnetism**.
Biosphere	That region of the earth's surface and atmosphere within which all living things are found.
Biotic Biotic factors	The **synergy** which exists between **nucleic acids** and **proteins**, sufficient to influence the natural environment and other organisms. See also, **biogeochemical norms**.
Bistable multivibrator/s (computer science)	In electronics, a bistable multivibrator has two stable states, enabling this ingenious piece of circuitry to be switched from a 'zero' to a 'one', and *vice versa*. Thus, computer memory exists in **binary notation** (making it compatible with **Boolean operators**, which process the same). See also, **unistable multivibrator** (neuroscience).
Black body (physics)	An imaginary object, according to orthodox physics, which is a perfect absorber and emitter in all **wavelengths**. The light and heat it emits are wholly dependent upon its temperature.
Black hole/s (dark sciences)	A strengthening **magnetic field** initiates the absorption of a **photon**, and a weakening **electric field** the **emission** of a photon; in the case of black holes, **gravitational** collapse (and hence escalating **fusion**) leads to a runaway strengthening of the 'singularity's' very own magnetic and electric fields. Theoretically, the black hole becomes a virtual **black body**, whose burgeoning magnetism makes it a perfect absorber (not simply of incident radiation, but also of **energy** emanating from its own core). Whether our galaxy's supermassive black hole gravitationally weakens with age, loosening its 'grip' on the milky way's spiralling arms, remains an open question. See also, **deep-space nucleosynthesis** and **stellar nucleosynthesis**.
Body Body cycle (human biology)	**Behaviourists** might analyze the body cycle, or **NE-cycle**, but it would require **cognitive psychologists** to deconstruct the **mind cycle**, and hence the **CN-cycle** (**personality** and **behaviour**, and thus **disposition**, deriving from **positive-feedback** and **negative-feedback** respectively).

194

Bond-cleavage (chemistry)	Dissociation requires **energy**, i.e. a stronger **electromotive force** and **nucleomotive force** (making bond cleavage a reversible, non-spontaneous, **endergonic** reaction).
Bond formation (chemistry)	Electronegative atoms, in close proximity, spontaneously weaken one another's peripheral magnetic fields, and hence their respective electric fields, forcing their valence **electrons** into lower, more stable, molecular orbitals, with the emission of a **photon**. **ATP hydrolysis** is similarly spontaneous, or **exergonic**, enabling non-spontaneous forms of bond formation and **bond cleavage** to occur in living systems.
Bonding (chemistry)	See **covalent bonding** and **ionic bonding**.
Boolean operator/s	In computer science, logic gates, latches and switches (which rely on **binary inputs**) are classified as a Boolean operators. Which has its corollary in the biological sciences, where **neurons** as similarly reliant upon high and low voltages, or electrical potentials.
Boom-bust cycle/s (economics)	The business cycle is characterized by an expansion and contraction of the economy, arising out of the interplay of innovation and obsolescence, together with fluctuating **supply** and **demand**.
Bottom-up (sociology)	In sociology, bottom-up refers to **social action** (which is often motivated by **anti-exploitation bias**, financial hardship and duress).
bottom-up (neurology)	**Sensory awareness** (producing **afferent signals**, which **empiricists** place supreme weight and importance upon).
Brain/s (neuroscience)	Borrowing from the **triune brain**, the neomammalian complex and paleomammalian complex (**telencephalon**, with its all-important **limbic system**) accommodate the reptilian complex (**basal ganglia**, **diencephalon**, etc), as it is, figuratively-speaking, pushed-upwards (all clicking neatly together, like a neurological press-stud). At the focus of this union of **mind** and **body**, where the **conceptual** and **corporeal** gyrate, is a structure termed the thalamus, which relays incoming sensory information to pertinent parts of the brain, where it can be processed (making the brain an **electrochemical** hub, custom-built to integrate both **radiological** and **hormonal** activity).
Brain lateralization (anatomy)	Women's **brains** are less lateralized (their corpus callosum, which aids communication between the hemispheres, being comparatively larger).
Brain wave/s	More permanent than **thoughts**, reverberating brain waves, arising from **inactivation-emission**, produce a **multivibrational matrix** (the hypothesized basis of human **consciousness**). See also, **absolute refractory period**.
Brexit Brexiteers (politics)	In the **EU referendum** (23 June 2016) a slim majority of the **UK** electorate voted to leave the **European Union**, giving rise to the term 'Brexit'. The significance of a **single integrated economy** (in terms of social progress, regional stability and raised living standards) has spawned an impassioned debate regarding Britain's internal and external borders; borders which the labour-intensive **markets** would much prefer to ignore. The free movement of goods, services and people would have been better framed as 'the free movement of goods, capital and labour', in-so-far as labour is synonymous with profit, whereas people suggest increased public expenditure (that is to say, free movement enables the exportation of goods at a profit; a profit which couldn't be realized without the additional labour).

British values	Objectively-speaking, that which a person can, or could, 'get away with', within the jurisdiction of the **United kingdom** (not to be confused with, for example, Nigerian or American values, which pertain to practices which one can, or could, 'get away with', within the jurisdiction of Nigeria and America, etc). That is to say, all such values are vulnerable to **dumbing-down**.
Bruising *"prove-it"* attitude	Common to those guilty of apostasy, in respect of professed values, textbook protocols and due process. See also, **mitre paradox**.
C	
Calculus	Calculus, both linguistic and mathematical, computes through the expedient of infinitesimally fragmenting statements, gradients and areas. See also, **cruciform calculus**.
Cambrian explosion (evolutionary biology)	**Life** on earth was hindered, for much of the 3 billion years which preceded the Cambrian explosion, by harsh conditions, low oxygen levels and damaging radiation. Crucially, **bacteria** evolved which were specialists in **organic decomposition**, **nitrogen-fixation** and **oxygenating photosynthesis**. Making the engine, driving that explosive speciation, largely bacterial.
Capitalism Capitalist/s	Free-market capitalism stresses that the **state** should be independent of **economic** production, placing responsibility for wealth creation in private hands. Ironically, the free **market**, so prized by the **Right-wing**, blurs state boundaries (boundaries which are unintentionally hardened by centralized planning and the anti-capitalist **Left-wing**). Therefore, globalization cleverly exasperates anyone not occupying the **political middle ground**. See also, **socialism**.
Carbon cycle	Consumers exhale carbon dioxide (and later decompose when dead), enabling carbon to re-enter the environment; carbon which photosynthesizing producers convert into edible carbohydrates.
Cartesian dualism Cartesian soundness	**Mind-body duality** reflects the fact that the **conceptual** and **corporeal** are equivalent, but by no means equal (that is to say, the **mind** governs the **body**, just as 'E' shapes 'mc²'). Cartesian dualism concedes that this dichotomy exists, and that brain waves are the individual's most ineluctable truth. See also, **'cogito, ergo sum'**, **competitiveness** and E=mc².
Catabolic Catabolized	The process by which complex molecules are broken-down into simpler ones, releasing **energy**, e.g. catabolism of stored fat.
Catalysis Catalyst	Chemical processes which speed-up reaction times, but leave the catalysts unchanged by the reaction.
Cell Cell cycle Cell division	The cell cycle has three stages: 1) interphase, involving cell growth, **protein synthesis** and regular functioning; 2) **mitotic cell division**, in which two fresh nuclei, containing indistinguishable **genetic** material, are produced; and 3) cytokinesis, during which the **cytoplasm** divides, separating the two nuclei and creating two completely new cells.
Cell wall	Structural support surrounding the cells of plants, bacteria, algae and some archaea. See also, **plasma membrane**.
Censorship (politics/sociology)	Censorship denotes a restriction of free expression (but when 'free expression' isn't matched by sufficient information with which to judge online content or popular content encourages illegitimate imitations perhaps expression ought to be curtailed).

Centralization (political)	The **political middle-ground** is torn between centralization and **decentralization** (that is, central government control versus control by local authorities and/or the private sector).
Centre Centre-ground Centrist politics (political)	The **political** Centre separates, fragments and denies **power**, without recourse to **subnational**, **national** or **supranational** interventions (that is to say, multi-party politics, pragmatism and public debate all dominate political life, with a strong emphasis on personal freedom, liberal individualism and civil liberties). See also, **neocentrism**.
Centromere (genetics)	The fertilized human egg possesses 46 **chromosomes**. During **mitotic cell division** each chromosome shortens and thickens, revealing itself as 2 **chromatids** (joined at a centromere, like an 'X').
Cerebral cortex Cerebral hemispheres Cerebrum (human anatomy)	The cerebrum is the largest portion of the **brain**, comprising two cerebral hemispheres, each possessing four lobes (termed the frontal, parietal, temporal and occipital lobes). **Conceptual processing** takes place in the outermost grey matter, or cerebral cortex, comprising gyri and sulci (ridges and furrows). See also, **telencephalon**.
Chaos theory Chaotic system (mathematics)	A chaotic system is sensitive to its initial starting conditions, such that negligible changes at the start may produce sweeping disorder later (provided, that is, that **positive-feedback** ensues). Indeed, **CN-cycles** may produce sweeping sociological disorder, and in spite of natural laws preferring stable **communities**.
Chemical bonding	See **covalent bonding**.
Chemistry	Orthodox chemistry addresses available **energy** (and the chemical reactions arising from the same). See also, **physics**.
Chloroplasts	See **primary endosymbiosis**.
Cholesterol Cholesterol homeostasis	Cholesterol homeostasis is central to a properly functioning **cognitive-nervous-endocrine axis** (cholesterol metabolism synthesizing a variety of **antagonistic hormones**). See also, *de novo* **synthesis**.
Chomskyan **linguistics**	American linguist Noam Chomsky (1928-) postulated that young children have an innate predilection for syntactic structure, common to all natural languages. However, such a predilection does not, in itself, guarantee superlative **critical thinking skills**.
Chromatid/s (genetics)	**Semiconservative replication** produces two identical strands of **DNA**, i.e. two chromatids, joined at a **centromere**, like a letter 'X'.
Chromosomes (genetics)	Rod-shaped structures, found within the nucleus of a **eukaryotic cell**, containing the whole **genetic** code, or **genes**, of a given individual.
Cistron (genetics)	That portion of a **gene** which codes for a specific characteristic.
Cleaving (chemistry)	See **bond cleavage**.
Climatic factors	Climatological conditions affecting the natural environment and its prevailing life-forms. See also, **biogeochemical norms**.
Climax communities Climax community	**Stabilizing selection** is the dominant feature of climax communities, whose **differential mortality** works in the interests of established **phenotypes**. **Stabilizing fecundity**, arising through **coevolution**, favours established **neurotypes**. See also, **progressive selection** and **psychodynamical climax**.

197

Cline/s	A **genetic** sub-population, subject to **progressive selection** (involving profitable mutations, beneficial gene-flow and discriminating forms of mating). See also, **demes**.
Clinically unsound	A **sub-aequipine** psychological state, requiring pharmaceutical correction. See also, **clinical normality**.
Clinical normality	A drug-free physiological or neurophysiological state, not requiring pharmaceutical correction. See also, **clinically unsound**.
Closed system/s (physics/chemistry)	A closed system can only exchange heat and **energy** with its surroundings. See also, **open system** (chemistry/biology).
Closed system/s (sociology/politics)	Closed systems possess negligible social mobility (their **ascriptive practices** leading to sexual inequality, political violence, racial tension and religious tyranny). See also, **open system** (sociology/politics).
CN-cycle/s	See **cognitive-nervous cycle**.
CNE-axis	See **cognitive-nervous-endocrine axis**.
Codification (constitutional law)	Non-negotiable statutes, codes and constitutions, which purport to exert a distributional **power** by virtue of being written-down.
Coding (computer science)	Slang for computer programming. See also, **digital determinism**.
Codon (genetics)	A typical **protein** contains twenty different **amino acids**, each coded for by a triplet of bases, called a codon. See also, **deoxyribonucleic acid**.
Coevolution Coevolutionary dynamic Coevolving	Also called duel inheritance, coevolution arises when **biologically-produced** and **socially-constructed** factors feedback, affecting the attributes of subsequent generations (that feedback resulting in racial differences, allied to superficial **sex-linked** characteristics). **Cruciform calculus** enables an **epistemic community** to enumerate variables as diverse as **neurotype**, **pragmatics**, **personality**, **behaviour**, **structures** and **genotype** (in order to constructively shape **cognitive**, **nervous** and **endocrinal** responses).
Cogitation	Thinking, reasoning and deliberation. See also, **cognitive pragmatics** and **critical thinking skills**.
'*cogito, ergo sum*' (I think, therefore I am)	If one doubts everything which can be doubted, however improbable the deception, one is left with indubitable **cognitive**, **nervous** and **endocrinal** responses (and the telling use of phrases like "*almost certainly is*", to describe lesser truths). See also, **Abelard's dictum**.
Cognitive	See **radiological**.
Cognitive bias	Denotes humankind's predilection for partiality, prejudice and discrimination, there being an expanding list of such biases. See also, **humanism**.
Cognitive-nervous cycle (CN-cycle)	The cognitive-nervous cycle, or **mind cycle**, is a cycle of **positive-feedback**, built upon **radiotransmission** and **neurotransmission**. See also, **nervous-endocrine cycle** and **cognitive-nervous-endocrine axis**.
Cognitive-nervous-endocrine axis (CNE-axes)	The CNE-axis comprises two divergent cycles: 1) the **mind cycle**, or **CN-cycle** (built upon **positive-feedback**, and driven by **radiotransmission** and **neurotransmission**); and 2) the **body cycle**, or **NE-cycle** (built upon **negative-feedback**, and driven by neurotransmission and **arteriotransmission**). The CNE-axis determines a person's **neurotypic sub-classification**, i.e. whether they're **epipoliotical**, **hermeneutical**, **epivevatical** or **ameroliptical**. See also, **triune brain**.

198

Cognitive pragmatic competence Cognitive pragmatics	Cognitive pragmatics denotes how a person thinks, reasons and deliberates. Which, depending on the context, may be classified as **general cognitive pragmatics** (routine thinking) or **applied cognitive pragmatics** (professional deliberation). Cognitive pragmatic competence denotes the propensity of one individual to accurately decode another's **mind**, when remotely bound together.
Cognitive psychologists Cognitive psychology	Cognitive psychologists have a **rationalist** perspective, which interprets **behaviour** as symptomatic of outgoing **efferent signals**, emerging from subjective **cogitation**. Unlike, **behaviourists**, who are **empirical**, believing that incoming **afferent signals** condition one's behaviour, or otherwise explain the same.
Cognitive system	See **radiological**.
Cold logic	Cold logic amounts to endorsing personal autonomy, sufficient to objectively appraise the **genetic** substructure.
Cold War (1946-1989)	The state of **political** and **military** tension which once existed between **The West** and Soviet Union. See also, **Iran-Saud-Arabia 'Cold War'**.
Collective power	Whereby a group gets others to carry out their combined will.
Colloquial Colloquialism	Commonplace language and unremarkable **cogitation**. See also, **exceptionality** and **informality**.
Communication systems (biology)	According to convention, the human **body** possesses two major communication systems: one **nervous**, or **electrochemical**, the other **endocrinal**, or **hormonal**. This exposition introduces a third – namely, the **cognitive**, or **radiological**. Together, these three communication systems use **neurotransmission**, **arteriotransmission** and **radiotransmission** respectively. See also, **cognitive-nervous-endocrine axis** and **telepathically-induced effects**.
Communist communism Communist Manifesto	Karl Marx's (1818-83) Communist Manifesto, published in 1848, concerns class struggle and "the property question", i.e. **capitalism** is exploitative, a stateless society essential, the reforming alternatives deficient, with working men needing to unite (the resultant 'classless' state conforming to a **pecuniary model of extremism**).
Communities *Community*	Stable **ecosystems** comprise competitive **species**, rather than competing ones – although individuals of the same **species** frequently compete with one another, thereby maintaining that **competitiveness**. *Homo sapiens* competes within itself, but not to be **competitive** within a stable **biosphere**.
Comparative advantage (politics/economics)	**Sapient** individuals, by virtue of **unsound induction**, possess, at best, only a comparative advantage (whereas **aequipine** individuals enjoy an **absolute advantage**, due to **sound deduction**). See also, **sapient economics** and **aequipine economics**.
Competing *'competing'*	Some see boys as *'competing'* (notwithstanding that the *'competitive'* aren't given to competing, and the act of competing doesn't prove **competitiveness**).
Competitive *'competitive'*	Some see girls as *'competitive'* (notwithstanding that *'competing'* isn't competitive, and **competitiveness** isn't actually appraised by competition).
Competitive exclusion	Grown independently, in a restricted space with limited resources, two closely-related **species** often exhibit **'s' shaped growth curves**. Grown together, however, the least **competitive** is eliminated. See also, **'one true' competitive exclusion**.

Competitiveness	Competitiveness isn't determined by humans, it's a **conceptual** judgement arising beyond ourselves, sufficient to decide which **species** of human inhabits our global niche. **Cruciform calculus** is an attempt at pre-empting that verdict. See also, **Almighty influence**.
Complex adaptive system/s Complexity theory (mathematics)	Complex adaptive systems, particularly those governed by **positive-feedback**, tend towards criticality, emergent properties and catastrophe (nascent patterns, emergent order and **negative-feedback** notwithstanding). See also, **acute non-linear event**.
Compound/s (chemistry)	A chemical compound is formed by two or more **elements** combining in definite proportions, the bonds being either **covalent** or **ionic**.
Concentrate Concentration of power (politics) Concentration of capital (economics)	The hunger for **power** (**economic, political, military** and **ideological**) is often emotionally driven, as evidenced by the horrors facing those who attempt to decompose the same. European imperialism was arguably the apogee, as regards the concentration of capital. See also, **decomposition of power** and **decomposition of capital**.
Concept/s Conceptual Conceptual framework Conceptual influences	**Conceptual reality**, like the human **mind**, comprises **electric** and **magnetic** fields. And, like the human mind, those astrophysical fields are capable of eclipsing the bodily, or corporeal, becoming the primary cause of events, rather than mere byproducts of the same (a kind of cosmological **consciousness**, or **Almighty influence**, which initiates rather than reacts). See also, **meme**.
Conceptualism (mathematics)	The view that outward reality resists algebraic expression, given that science relies on **probability**, **extrapolation** and **statistics**.
Conceptualism (philosophy)	Vector interpretation of reality, which soundly ascertains the strength and direction of the **e.m.f** and **n.m.f**. See also, **corporealism**.
Conceptual processing (psychology)	The **mind cycle** integrates *bottom-up* **sensory awareness** (**empirical** observation) with *top-down* conceptual processing (**rationalist** deliberation). See also, **'a priori' coniunctum 'a posteriori'**.
Conceptual-QED (dark sciences)	One of the four **sub-atomic quadrants** making-up the **dark cycle atomic model** (concerned with the **electric field**, **electromagnetic induction** and the **electromotive force**). See also, **corporeal-QED**.
Conceptual-QSD (dark sciences)	One of the four **sub-atomic quadrants** making-up the **dark cycle atomic model** (concerned with the **magnetic field**, **electromagnetic induction** and the **nucleomotive force**). See also, **corporeal-QSD**.
Conceptual reality Conceptual states	Conceptual reality comprises **electric** and **magnetic** fields (together with **electromagnetism, electromagnetic induction, electromotive forces, nucleomotive forces** and the **laws of thermodynamics**). Accordingly, **biased, logical, sound** and **balanced** minds, like temperature, are conceptual states. As such, **scientific** research informs conceptual reality, by making erudite predictions about **corporeal reality**, which may, or may not, withstand experimental testing.
Configurational determinism Configurational theory (dark sciences)	Configurational theory argues that sub-atomic forces replicate themselves at all scales, skewing **evolution** and providing for a **multivibrational matrix** (a case of the primacy of the **electric** and **magnetic** fields being conserved). See also, $E=mc^2$.
Connectives	See **logical connectives**.

Connotation Connote/s	Beyond **denotation**, one encounters the feelings and ideas which are commonly associated with a lexeme, phrase or concept.
Connubial dualism (neuro-cognitive effects)	An ethical utilization of **telepathically-induced effects**, arising when two minds develop together as one convergent **mind cycle**. See also, **heteronnubial dualism** and **homonnubial dualism**.
Consanguineous collective/s (anthropology)	Originally humans organized themselves into clans, tribes and chiefdoms, i.e. consanguineous collectives, founded upon common ancestry. Later, the **functionally-nucleated family** arose, thanks to its compatibility with highly-developed economies.
Consciousness	After the **mind cycle** has been running for several years, the sheer volume of **thoughts** makes anything other than an overshadowing sense of self impossible (if replicated digitally, a computer would become conscious of its own existence). This anthropic property arises because **radiotransmission** has stopped being a by-product of **neurotransmission**, and has become, instead, the principal cause of the same. See also, **multivibrational psychometrics** and **conceptual influences**.
Conservation Conservation law (physics/dark sciences)	For every conservation law, there's a **symmetry** (a symmetry arising whenever a system undergoes a substantive change, but aspects of it remain unaltered). The **fourth law of thermodynamics** states that **electrons** and **protons** respond predictably to changes in the **electric** and **magnetic** fields, because those particles possess an **electronic constant** and **nucleonic constant** respectively, i.e. conserved characteristics.
Conservation of electronics (dark sciences)	The writer's **fourth law of thermodynamics** states: *"the electric and magnetic fields, acting upon charged particles which are in motion, must be of a prescribed intensity, otherwise the said particles will jump or fall into different **orbitals**, or else realign in the case of atomic nuclei"*. Accordingly, when an atom's **electric field** weakens, an **electron** must fall into a lower-energy orbital, in order to experience the stipulated field strength (with **magnetic field lines** creating variable electric fields, consistent with shells, subshells and orbitals).
Conservation of energy (chemistry)	**Energy** can neither be created nor destroyed (otherwise known as the **first law of thermodynamics**).
Conservation of mass (physics)	Matter can neither be created nor destroyed. See also, **deep-space nucleosynthesis** and **stellar nucleosynthesis**.
Constants (linguistics)	See **logical constants**.
Constructive interference (physics)	Occurs when two identical waves are perfectly in phase, sufficient to amplify one another. See also, **synergy**.
Constructive theory of life's origins (evolutionary theory)	A life-sustaining **synergy** exists between **nucleic acids** and **proteins** (prior to which, there was only **evolution** and **extinction**, accreting **pre-biotic** detritus). This book's constructive theory of life's origins likens it to striking a flint onto a combustible material (that is, synergize correctly and **life** explodes across a nutrient-rich planet). See also, **biological singularity**.
Constructivism (mathematics)	Constructivist mathematics is intuitionist, with algebra perceived as a mental construct, incorporating internally consistent methods and contrived mathematical objects.

Contiguous	Adjacent to, but separate from, e.g. **neurons**.
Conventional current flow (electronics)	Conventional current flow runs from the positive to negative terminal (with **real electron flow** running from the negative to positive terminal). **Configurational theory** doesn't accept that real or conventional currents generate **magnetic fields**, as this would imply that particles have primacy over fields. Instead, it argues that a **potential difference** arises due to changes in the **electric field** (which has the appearance of a current generating a magnetic field).
Corporeal	See **corporeal reality**.
Corporealism (philosophy)	Materialistic **interpretation** of reality, which **logically** supports an expanding list of particles. See also, **conceptualism**.
Corporeal-QED (dark sciences)	One of the four **sub-atomic quadrants** making-up the **dark cycle atomic model** (concerned with **electrons, electrostatic attraction** and the **electromotive force**). See also, **conceptual-QED**.
Corporeal-QSD (dark sciences)	One of the four **sub-atomic quadrants** making-up the **dark cycle atomic model** (concerned with **nucleons, electrostatic attraction** and the **nucleomotive force**). See also, **conceptual-QSD**.
Corporeal reality Corporeal states	Comprises **protons, neutrons** and **electrons** (together with **gravitational forces, dispersional forces, spatialgravitism, electrostatic attraction** and the **laws of topodynamics**). The **dark sciences** define **solidity** as the mutual repulsion of electrons (making gases, liquids and vapours all technically solids). See also, **conceptual reality**.
Cosmic background radiation (CMBR)	The CMBR comprises radiation in a bandwidth between the visible and radiofrequency wavelengths, which is discernable from every direction (which the **dark sciences** interpret as sufficient **mass** forming, in the universe's past, to support a surfeit of **energy**). See also, $E=mc^2$.
Coulomb's law (classical physics)	Coulomb's law states that the **electric** force of attraction is proportional to the product of the charges, and inversely proportional to the square of the distance between them. See also, **Maxwell's equations**.
Covalent covalent bonding (chemistry)	Covalent bonds form whenever two atoms share an **electron**, thereby achieving full **valence shells**, increased stability and a lower-energy state. See also **molecular bond-state**.
CRH	See **corticotropin-releasing hormone**.
Criminal solicitation	Encouraging criminal misconduct or contributing to the same.
Critical thining skills	Denotes a person's ability to construct arguments, critically evaluate the same and assess the quality of supporting evidence. **Prescriptive reasoning** is **epivevatical** (that is, scientific deliberation) or **ameroliptical** (whereby the context, within which such knowledge is soundly amassed, is objectively appraised, and the facts responded to in a properly **well-balanced** manner). See also, **cognitive pragmatics** and **neurotypic sub-classifications**.
Crossing over (genetics)	During **meiosis** identical strands of **DNA** are produced (that is to say, each **chromosome** becomes two **chromatids**, joined at a **centromere**, like a letter 'X'). Homologous chromosomes (comprising two chromatids apiece, joined at centromeres) can then exchange small segments of **genetic** material, in a manner known as crossing over). See also, **independent assortment**.
Crucible of social action	See **social action crucible**.

Cruciform calculus Cruciform coercion (multivariate analysis)	Cruciform **calculus** deconstructs the **coevolutionary dynamic**, using axes which enumerate **biologically-produced** and **socially-constructed** factors (the sum of the biological-produced factors indicating the **agent** or **neurotype**, as marked on the **'y' axis**, and the sum of the socially-constructed factors indicating the **structure** or **sociopolitical superstructure**, as marked on the **'x' axis**). See also, **x-y coordinate system**.
Cruciform coevolution	The **biosphere** comprises countless systems – ones married together prehistorically by **ecology**, but historically, by psychology. However, whilst human psychology continues to supplant ecology, anthropoid psychology remains prey to evolutionary laws. In other words, *Hominini* jeopardizes whole ecosystems, whereas cruciform coevolution raises the rationale of humans, sufficient to benefit both the environment and the people inhabiting the same. The academic study of humankind's relationship to the biosphere is termed **anthropobionomics**, with **cruciform calculus** steering humanity towards a rapport which is competitively **aequipine**, rather than competingly **sapient**. Without question, the limits of cruciform coevolution are delineated by **structurating** agents and **socializing** structures, beyond which one meets with a mandatory humility arising from the ramifications of **configurational determinism**. See also, **unnatural fiscal selection** and **cruciform laws**.
Cruciform laws Cruciform model Cruciform sociometry	To avoid becoming crucibles for self-defeating **sapiency**, countries must submit to cruciform laws, i.e. inflexible laws administered by means of **cruciform calculus** (using data from cruciform sociometry).
Currency wars (economics)	Currency warring is a temptation, particularly amongst Right-wing **monetarists**, keen to boost foreign trade by devaluing their currency. A ploy made easier, of course, if one's currency is independent, like the country itself. See also, **Right-wing**, **deflation** and **Brexit**.
Cybernetics (artificial intelligence)	Cybernetics denotes an **electronic** system with an in-built capacity for **homeostasis** and/or **operant conditioning** (which may involve a robot shaping its own **personality** and **behaviour**, in response to external stimuli, or an evaluation of **biogeochemical norms**, in order to shape humanity's personality and behaviour *en masse*).
Cytoplasm (biology)	The gel-like substance found within cells, which occupies the space between the **nucleus** and the **plasma membrane**.
D	
Dark cycle atomic model (dark sciences)	Where incident radiation strengthens an atom's **magnetic field**, **electromagnetic induction** will simultaneously strengthen the associated **electric field**, forcing an **electron** to jump-up into a higher-energy **orbital**, as a **photon** is absorbed. In effect, the magnetic field realigns the **nucleus**, having generated a **nucleomotive force**, thereby weakening the nucleus' **electrostatic attraction** to the said electron (the electric field simultaneously exerting an **electromotive force** on the electron). If the electric field subsequently weakens, a photon will be emitted as the electron falls-back into its **ground-state**, with the nucleus returning to its original alignment. Implicit in this argument is the notion that electrostatic attraction is inversely proportional to **energy** (such that **neutrons** form if energy is absent).

Dark cycle theory of everything (dark sciences)	Gravitational forces conform to the **inverse square law**, such that their impact radiates outwards, becoming ever weaker. So weak, say the **dark sciences**, that **gravity** becomes **gravitation**, inhibiting the passage of **light** (that light breaking-down into charged particles, the **electric field** creating **electrons** and the **magnetic field** creating **protons**, with any **neutrons** formed being vulnerable to **beta decay**). In addition to creating particles, this **deep-space nucleosynthesis** also fabricates supplementary **gravitational** and **dispersional** fields. At the other gravitational extreme, **stellar nucleosynthesis**, arising from massive concentrations of particles, radiates electric and magnetic fields. The above fields and forces, i.e. **electromagnetism** and **spatialgravitism**, comprise isolines and gradients; which can be examined using **tensor-poiesis**, so as to better understand how the **electronic constant** and **nucleonic constant** cause particles to move within the fabric of **space-time**.
Dark energy (dark sciences)	The **energy** associated with the conversion of **electromagnetic radiation** into **mass-dispersion**. See also, **dark matter**.
Dark matter (dark sciences)	Objects rendered invisible by virtue of their emitted or reflected **electromagnetic radiation** converting into **mass-dispersion**.
Dark sciences	Sciences founded upon the **dark cycle atomic model** and **dark cycle theory of everything** (models which have implications in every scientific field). The dark sciences, ever conscious of classical physics' longstanding problems of extrapolating from fragmentary experimental data, shun work from the general to the particular (that is, from comprehensive theory to detailed mathematics, by way of experimentation). See also, **Maxwell's equations**.
Debiasing (cruciform advance)	**Neocentrism** counters the **top-down** weakness for despotism and the **bottom-up** predilection for disorder, by coevolving **neurotypes** less prone to **cognitive bias**.
Decentralization (political)	A transfer of central government control to local authorities and/or the private sector (with the **political middle-ground** torn between **centralization** and decentralization).
Declarative **knowledge**	Established or purported facts, figures and statistics. See also, **procedural knowledge**.
Decomposition of **capital** (economics)	Process by which private ownership diversifies, amid the class system's terminal collapse. See also, **decomposition of power**
Decomposition of **power** (political)	A decomposition of **economic, political, military** and **ideological** power is achieved through mainstream human agency, as per **cruciform calculus** (in the case of economic power, a **decomposition of capital** is often called-for). **Absolute prosperity** can only be achieved through the decomposition of power (with **absolute poverty** being alleviated through **socialism's** decomposition of capital).
Deduction Deductive	See **hypothetico-deductive cycle**.
Deep-space **nucleosynthesis**	See **dark cycle theory of everything**.
Deflation (economics)	Prior to World War Two, deflation and **inflation** worked together in tandem, keeping wages and prices stable. That is, deflation increased the value of a currency, as wages and prices fell, triggering inflation-inducing demand, followed by a reduction in the currency's value (in other words, deflation makes one's money go further and inflation makes it go less far). After the war, **economic interventions** confined deflation largely to history, whereupon wages and prices slowly crept-up, as the **Left-wing** and **Right-wing** vied to reign-in inflation and control **unemployment**. See also, **stagflation**.

Defragment (politics)	In the case of social disorder, one must defragment those elements which combine to produce **hegemonic democracy** (beginning with the **economic** sector). See also, **unnatural fiscal selection**.
Deleterious Deleterious community	Communities which fail to discriminate against deviations from the **political middle-ground**, and where fecundity is neither stabilizing nor progressive in respect of **balanced-mindedness**.
Demand (economics)	See **aggregate demand**.
Deme/s	Demes are genetically isolated sub-populations, subject to **stabilizing selection**. *Homo aequipondium amerolipticus* may one day rank as a deme (subject to **stabilizing fecundity** within a **global governance complex**). All other **neurotypic sub-classifications** behaving like **clines**, subject to **progressive selection** (eventually rendering them **well-balanced** over the course of geological time).
Democracy	See **hegemonic democracy**.
Dendrital output (dark sciences)	Provided a signal has been received via an **axonal input**, a **unstable multivibrator** will tirelessly reverberate due to **neuronal induction**. To access the same, a **neuron**, serving as a dendrital output, must first be stimulated with a different frequency, whereupon the conscious **mind** can establish whether a 'one' or a 'zero' has been saved. See also, **memories** and **Boolean operator**.
Denotation Denote/s, Denoting	A word or phrase's literal meaning.
de novo **synthesis**	The **brain** is incapable of absorbing **cholesterol** directly from the blood stream, due to the 'blood-brain-barrier' – therefore, it must synthesize cholesterol *in situ* using far simpler molecules, in a process termed *de novo* synthesis.
Denunciation/s	Denunciation is an aspect of **Right-wing scapegoating**, as experienced by the widower of murdered Labour MP, Jo Cox (1974-2016). In addition to treating individuals as '*guilty, until proven innocent*', the **Right-wing** also ignorantly revels in the taking of human life, due to its penchant for **dispositional attribution** (a predilection which obscures deeper and more complex sociological factors).
Deoxyribonucleic acid (DNA)	DNA contains the **genetic** information for a given individual. Each strand comprises enormous numbers of nucleotides, each possessing a nitrogenous base, a sugar and a phosphate group (the sugar of one being bonded to the phosphate group of the next, and so on). Bases come in four varieties, i.e. guanine (G), cytosine (C), adenine (A) and thymine (T). However, complimentary strands of DNA, bound together in a double-helix, have only the following base-pairings, A-T and C-G.
Dependent variable/s (mathematics)	A value which is contingent upon an **independent variable**.
Depolarization (neurology)	At rest, the membrane of a **neuron** is slightly polarized, with the inside negatively-charged. Outside the membrane are positively-charged sodium ions, which can pass through the membrane rapidly, via **sodium ion channels**, when the neuron is sufficiently stimulated (that depolarization resulting in a nerve impulse).
Deradicalization	Moderation of extremist thinking, initiated by those occupying the **political middle-ground**. However, deradicalization (like **socialization**, **transculturation** and **acculturation**) cannot revise a person's **CNE-axis**. In that sense, it's a form of sublimation.

205

Destructive interference, 2014 (Mark Fox)	Destructive Interference: Understanding the brain's telepathic potential is an account of how the human brain's telepathic potential was ingeniously unlocked between the years **AD 1913-47**. ISBN: 978-0-9929796-1-4 (paperback version). Also available as a Kindle e-book.
Deterioration	Occurs when a word or phrase assumes a more negative meaning.
Deviance (linguistic/cognitive)	A living language experiences phonetic, lexical and grammatical deviance. As regards **cogitation**, **cognitive pragmatics** addresses **multi-dimensional deviance**, whereby liberties are also taken with argument, proof and reason.
Devolution	A form of decentralization, whereby powers are passed down to regional assemblies and local parliaments (**sovereignty**, however, remains with that doing the devolving).
D.f (dispersive fields)	Strong **dispersional fields**, occupying the **intergalactic medium**.
d.f (dispersion forces)	Weak **dispersional fields**, occupying the **galactic medium**.
Diegesis	Plan, storyline or plot.
Diencephalon (forebrain)	The diencephalon is the most antiquated part of the **forebrain**, and includes the thalamus and hypothalamus (which, together with the more advanced **basal ganglia**, is the **NE-cycle**'s control centre, or 'brain', as regards **negative-feedback**). One proposes reclassifying the basal ganglia, so that it forms part of the diencephalon – this anatomical reclassification reflecting the fact that the entire NE-cycle behaves like a pre-mammalian central nervous system.
Differential calculus	The branch of **calculus** which analyzes rates of change.
Differential fecundity	Whenever a technologically advanced society, subject to **coevolution**, undergoes an **epidemiological transition**, **differential mortality** is replaced by differential fecundity.
Differential mortality	Differential mortality may discriminate in favour of established **phenotypes**, or against them. That is, **evolution** is governed by differential mortality and **coevolution** by **differential fecundity**.
Differentiated cell Differentiation (genetics)	When a spermatozoon and ovum fuse, forming a **zygote**, cell division produces a child possessing a complex anatomy. Differentiation is the mechanism which switches **genes** on and off, sufficient to account for those cellular differences. Arguably, differentiating between genes of differing function owes something to differentiating between ones of comparable function. See also, **dominant** and **recessive**.
Diffuse power	Compliance without recourse to express commands.
Diffusion (cellular biology)	Occurs when molecules diffuse across a **plasma membrane**, from a higher concentration to a lower concentration.
Digital determinism (1947–)	Digital determinism reflects humanity's growing reliance on **Boolean operators**. Expanding upon the same, software engineers, unable to replicate the brilliance of the early digital pioneers, are capable of marring progress – as evidenced by those who find themselves counselling one another online, due to deficiencies in previously ground-breaking programs. See also, **dumbing down**.
Direct square law (Classical physics)	Arises when a force's impact increases fourfold with each doubling of the distance. In classical terms, the **dispersional force** conforms to the direct square law, in-so-far as a diminution of **gravitational forces** provides for its escalating impact. See also, **inverse square law** and **Maxwell's equations**.

206

Dispersion	See **dispersion force**.
Dispersional Dispersional field/s Dispersional force/s (dark sciences)	Dispersional fields (be it **dispersion forces** or **dispersive fields**) provide for **solidity**, due to the mutual repulsion of **electrons**. The **intergalactic** generation of dispersional fields is inflationary, due to **gravitational** forces building gradually. Adjacent neutrons experience a **weak nuclear force**, due to them containing **protons** and electrons. See also, **spectrum of repulsion** and **direct square law**.
Dispersion force/s (d.f)	The **dispersional field**, as found within the **galactic medium**. Molecules in close proximity may experience mild **electrostatic attraction**, due to their **electrons** repelling one another.
Dispersive field/s (D.f)	The **dispersional field**, as found within the **intergalactic medium**. It is the cumulative impact of these fields which drives cosmic inflation.
Disposition	The sum of one's **personality** and **behaviour**, save-and-except for the distorting impact of circumstance. See also, **dispositional attribution**, **situational attribution** and **mitre paradox**.
Dispositional attribution (psychology)	Arises when **behaviour** is attributed to a person's **disposition**, rather than the demands of the situation. The **Right-wing** is inclined towards dispositional attribution, preferring to blame the individual, rather than the prevailing conditions (in the case of capital punishment, Right-wingers glory in the taking of human life, whilst ignoring the critical impact of environment, including **radiofrequency technologies**). See also, **situational attribution**.
Disaggregated state/s (politics)	A **state** which possesses a **separation of powers**, **multilevel governance** and **coevolutionary dynamic**.
Distributional power	Arises when an individual gets others to carry out their will.
DNA	See **deoxyribonucleic acid**.
Domains	See **three domains**.
Dominant (genetics)	Different versions of the same **gene** are called **alleles**, with some being dominant and therefore expressed, even in a **heterozygous** form. Dominance may reflect the ease with which **functioning proteins** are synthesized, or **epistasis** may cause monitor genes to switch alleles between a dominant and recessive state. Alternatively, genes of comparable function may also meet with **differentiation**.
Double-blind controls	A scientific methodology which prevents **cognitive bias** invalidating a result (the experimenter not being aware of the working **hypothesis**). See also, **peer review**.
Double recessive (genetics)	A recessive gene can only express itself, as a **phenotype** or **neurotype**, in its **homozygous** form (known as a double recessive).
'duality of gender' Duality of structure (sociology)	The **coevolutionary dynamic** comprises a duality of structure, whereby **biologically-produced** agents and **socially-constructed** structures mutually-influence, by means of **structuration** and **socialization** respectively. **Sex** and **gender** form part of this dynamic, with society's **gender dialectic** influencing **sex-linked** characteristics, and *vice versa*.
Duel inheritance	See **coevolutionary dynamic**.
Due process (law)	Due process demands that a defendant's legal rights be respected.

Dumbing-down (sociology/law)	The body politic comprises a broad spectrum of **neuro-types**, leading to corrosive **social learning**. That is to say, **balanced-minded** protocols are vulnerable to peer-to-peer communication, of a kind which undermines the **scientific method**, **prescriptive reasoning** and **due process**. See also, **British values**, **digital determinism** and **intelligence**.
Dynamic equilibrium (chemistry/biology)	In the context of chemistry, a **closed system**'s reversible **equilibrium reactions** maintain chemical concentrations in a steady-state. In its generic biological form, an **open system** uses available **energy** to maintain a steady-state. In that sense, life arose as an open system, bathed in available energy, which sought to be impervious to change.
Dynamics (physics)	The **science** of how forces produce motion.
E	
e⁻	See **electron**.
Ecological niche/s Ecology	An ecological niche is the place occupied by a **species** within a **community**. See also, **trophic level** and **Gause's principle**.
Economic (power)	Commonly defined as GDP, i.e. the market value of all goods and services produced in one year (placing the **USA** in first place, followed by the **EU** and China). **Ideological**, **military** and **political** powers are contingent upon economic power, creating distrust in **globalization**. See also, **currency wars** and **global financial system**.
Economic equilibrium	A cruciform **global financial system** renders radical econo-mies disordered, rather than having the 'ecological whole' dis-ordered by feckless radicals. Political pathologies, engaged in **unsound induction**, arrive at unstable economies pos-sessing toxic levels of **positive-feedback**. The **political mid-dle-ground**, however, enjoys economic equilibrium, by virtue of **negative-feedback** and **sound deduction**.
Economic growth Economic infrastructure	The **economic** infrastructure supports the **sociopolitical superstructure**. In peacetime, fiscal policies seek to grow an economy, but to fully emulate the West German 'economic miracle' would require a rejection of **political**, **military** and **ideological** distractions, sufficient to invite the confidence of foreign investors.
Economic intervention/s	Economic interventions are used to steer advanced econo-mies, either through **fiscal policies**, related to **tax revenue** and **government expenditure**, or through **monetary poli-cies**, involving the national bank's management of **interest rates** and **money supply**. Customarily, such interventions have sought to raise or reduce **aggregate demand**, thus avoiding underproduction or overproduction.
Ecosystem/s	Self-contained ecological communities, comprising plants and animals, which interact with one another and with the natural environment.
Edaphic factors	Factors pertaining to the soils, which influence the natural environment and the organisms which inhabit the same.
E.f	Strong **electric fields**, occupying the **galactic medium**.
e.f	Weak **electric fields**, occupying the **intergalactic medium**.
Efferent signals (neurology)	Nerve impulses which are conveyed **electrochemically** away from the **brain**. See also, **rationalist**.

Ego (Freudian psychology)	Synonymous with one's self-schema, but embraces **consciousness**, **memory** and **Chomskyan linguistics**. See also, **id** and **superego**.
Electric Electrically Electric field/s Electricity (dark sciences)	Electric fields break-down into **electrons** and **dispersional forces** in the presence of **gravitation** (with **magnetic fields** breaking-down into **protons** and **gravitational forces**). **Electrostatic attraction** between protons and electrons is inversely proportional to the electric and magnetic fields' strengths, i.e. attraction is greatest when those fields are weak. Thus, hydrogen gas spontaneously 'precipitates-out' following **deep-space nucleosynthesis**. To weaken that attraction, enabling electricity to flow, requires a surfeit of energy (termed a **potential difference**). See also, **configurational theory**.
Electrochemical (nervous system)	One of the human body's three major communication systems. See also, **hormonal**, **radiological** and **electrochemically**.
Electrochemically Electrochemical signals	Nerve impulses conveyed by means of the **nervous** system. The electrical potentials associated **contiguous** neurons are 1. resting potential, 2. local potential, 3. threshold potential, and 4. action potential. The action potential is synonymous with a **nerve impulse** being triggered. See also, **neurons** and **electrochemical**.
Electromagnetic induction (dark sciences)	A moving or changing **magnetic field** has the appearance of inducing an electrical current in a current-carrying wire. According to the **dark cycle atomic model** any such fluctuation immediately impacts upon the **electric field**, exerting an **electromotive force** on the wire's **electrons** (whilst simultaneously exerting a **nucleomotive force** on the wire's atomic nuclei). See also, **electricity**.
Electromagnetic radiation Electromagnetic spectrum	**Energy** emitted by a light source, which disseminates according to the **inverse square law** (and which, depending on one's **interpretation**, takes the form of a **photon**, **transverse wave** or oscillating **electric** and **magnetic** fields). For electromagnetic radiation to propagate, **gravity** is required, which is a function of **mass-dispersion**; with dispersional forces dictating the **speed of light** (hence mc^2). The electromagnetic spectrum (spanning the range from harmful gamma radiation and X-rays, through to radio waves and extra-low frequencies) 'fills-up' due to the activities of the **galactic medium**.
Electromagnetism (dark sciences)	Electric and **magnetic** fields interact as electromagnetism, in the same way that **gravitational** and **dispersional** fields interact as **spatialgravitism**. Electromagnetism exerts **electromotive** and **nucleomotive** forces, giving rise to **phases of matter**.
Electromotive Electromotive force/s (dark sciences)	The electromotive force (e.m.f) arises as a consequence of the **fourth law of thermodynamics** (being one of three forces acting on **electrons**, the other two being **dispersional** and **electrostatic**).
Electronegative Electronegativity	Measure of how strongly an atom of a given **element** attracts an **electron**, sufficient to fill a partially-empty **valence shell**.
Electronic configuration	The electronic configuration of helium is $1s^2$ (that is, two **electrons** occupy an atom's 's' orbital, in shell 1). The electronic configuration utilizes inflexible **physics** to produce flexible **chemistry**, and hence **phases of matter** (including material properties such as viscosity, ductility, malleability and elasticity). See also, **configurational theory**.

Electronic constant/s	If one thinks of the **electric** and **dispersional** fields as comprising isolines and gradients, then the electronic constant forces an **electron** to move under their influence. See also, **electrostatic attraction**.
Electronic/s	The harnessing of moving **electrons**, primarily around electrical circuits, but also across vacuums and through gaseous media.
Electron jump/s (chemistry)	An **electron** either absorbs or emits a **photon**, jumping-up into a higher orbital during **absorption** and falling down into a **ground-state** during **emission**. See also, **molecular bond state**.
Electron/s (e⁻)	Particles which carry a single **negative charge**, but negligible **mass**. See also, **electricity** and **electron jumps**.
Electron volts (eV)	The electron volt is the kinetic energy gained by an **electron** as it accelerates through a **potential difference** of one volt, under the influence of an **electromotive force** (there being a comparable gain in energy associated with the **nucleomotive force**, hence **magnetic resonance imaging**, which energizes carbon and hydrogen atoms).
Electrostatic Electrostatically Electrostatic attraction Electrostatics (dark sciences)	Electrostatic attraction increases as **energy** wanes. The **dark sciences** arguing that *distance* (in coulomb's calculation for electrostatic attraction) should be replaced by *energy*, since *distance* is an **illusory correlation**. Building on that argument, the loss of an **electron** turns an **atom** into an ion, a particle noted for its electrostatic attraction (the atom's capacity for retaining energy having been diminished). Electrostatic attraction is to **spatialgravitism**, what **electromagnetic induction** is to **electromagnetism**. See also, **positive** and **negative**.
Element/s (chemistry)	Atoms of the same element contain identical numbers of **protons** (the earth having 92 naturally-occurring elements).
Emancipation	Equality and social justice, arising in the **disaggregated state**.
E=mc² (where '=' means equivalent, but by no means equal to)	Building upon Albert Einstein's (1879-1955) celebrated equation E=mc², this book argues that E (**energy**) and mc² (**mass-dispersion**) are equivalent, but by no means equal. That is, the former dominates the latter, with the **conceptual** shaping the **corporeal**. See also, **Life** and **Almighty influence**.
e.m.f	See **electromotive force**.
Emission (dark sciences)	A **photon** is emitted when an atom's **electric field** weakens, and an **electron** falls back into a **ground-state** (or, in the case of **chemical bonding**, a **molecular bond state**).
Emotion/s Emotional	**Socially-constructed** emotions are to **personality**, what biologically-produced **basal responses** are to **behaviour**. However, while emotions have an exaggerated effect on personality, due to the **CN-cycle's** use of **positive-feedback** (the **NE-cycle's** use of **negative-feedback** ensures that **biologically-produced** basal responses influence behaviour in relatively short bursts). See also, **disposition**.
Empirical Empiricism Empiricist/s	Empiricism sees knowledge as drawn from experience, via one's senses, whereas **rationalism** sees knowledge as the product of reason. Those who afford **afferent signals** more weight and importance than **efferent signals** are empirical. See also, modus operandi.
E=mv²/2 (physics)	The equation which calculates kinetic energy. It differs from E=mc², in-so-far as much more **energy** is released through the conversion of **mass**, than could ever be expressed by its acceleration.

Endergonic (energy-demanding)	Some molecules cannot be synthesized spontaneously, but must be metabolized within an organism (for example, the formation of sugars during **oxygenating photosynthesis** is one such endergonic reaction, requiring energy from sunlight). See also, **exergonic**.
Endocrinal Endocrine Endocrine system	See **hormonal**.
Endogamy	Only permitting marriage within a narrow social group, caste or clan.
Endogeny	Originating within society. See also, **exogeny**.
Endoplasmic reticulum	A type of **organelle**, found within **eukaryotic cells**, which comes in rough and smooth varieties (varieties associated with **protein synthesis** and steroid hormone production respectively).
Energy (E) (dark sciences)	Energy, in the form of **electromagnetic radiation**, is absorbed and emitted by **atoms** and molecules in response to a strengthening or weakening of their **electric** and **magnetic** fields.
Entropy (dark sciences)	According to the **second law of thermodynamics**, the entropy of an **isolated system** increases with time, leading to mounting disorder and a reduction in available **energy**. However, even if entropy does increase, resulting in the universe's heat death, **black holes** would still constitute an unstable concentration of unavailable energy (capable of extinguishing that otherwise unalleviated disorder).
Environmental resistance Environmental stress	All species meet with environmental resistance, due to **climatic factors**, **edaphic factors**, **oceanic factors**, **biotic factors** or **anthropogenic factors** (and, whilst **evolution** provides for organisms surmounting such resistance, some forms remain unsurmountable).
Enzyme/s	A specialized **protein**, synthesized by living organisms for the express purpose of speeding-up biological processes.
Epidemiological transition	Whereby mortality shifts from being a consequence of infectious disease, compounded by warring, to being one of degenerative illness, compounded by indulgence. See also, **differential fecundity**.
Epipoliotical Epipoliotical reasoning	Epipoliotical individuals possess poor ancillary skills, together with an excess of '*non-sequiturs*' (meaning that their conclusions rarely follow logically from their premises). Such reasoning is heavily reliant upon **ideological** assumptions (not least, the routine use of **ascription**). Additionally, the epipoliotical compensate with emotive language, most notably swearing, obloquy and invective). Which, in extremis, merges with the incoherent ramblings of the criminally insane. See also, **hermeneutical reasoning** and **proscriptive reasoning**.
Epipoliotics (scientific field)	The scientific study of **biased minds**, as regards their origin, impact and expression (from *epipolaiotis*, Greek for cursoriness).
Epistasis	More than one **gene** contributing to an expressed **phenotype** or **neurotype**.
Epistemic community	A supranational body of **well-balanced** experts, charged with overseeing the application of **radiofrequency technologies**. See also, **global governance complex** and **NEURON**.
Epistemological jurisprudence Epistemology	Epistemolgy is concerned with knowledge, truth and perception (with epistemological jurisprudence addressing what is legally knowable, or otherwise **admissible**, in an age of telepathic awareness).

Epivevatical Epivevatical reasoning	Epivevatical reasoning denotes the construction of **sound** arguments, by individuals possessing excellent ancillary skills. Additionally, those scientifically **sound minds** value first-rate source material, respected authorities and **peer review**. Accordingly, such reasoning contains all the necessary **propositions**, together with sufficient use of **logical constants**, to reliably convey facts. See also, **ameroliptical reasoning**, **prescriptive reasoning** and **critical thinking skills**.
Epivevatics (scientific field)	The scientific study of **sound minds**, as regards their origin, impact and expression (from *epivevaiosi*, Greek for confirmation).
Equilibrium reactions (chemistry/biology)	Equilibrium reactions are reversible in a **closed system** (that is to say, the **reactants** form **products**, and vice versa), enabling a **dynamic equilibrium** to be achieved. The cell is an **open system**, whose dynamic equilibrium is maintained with the help of **diffusion**, **osmosis** and **active transport**. In other words, the equilibrium reactions which take place in open and closed systems differ markedly in their scale, complexity and form. See also, **homeostasis**.
Ethnomethodology	**Microsociological** interpretation of society, which offers a linguistic explanation of how society is reproduced. See also, **social learning**.
EU	See **European Union**.
Eukaryota Eukaryote/s Eukaryotic cell/s	Eukaryota contains four kingdoms (Protista, Fungi, Plantae and Animalia). Eukaryote cells possess a discrete **nucleus**, containing the organism's **chromosomes**. Specialist eukaryotic cells have evolved which aid support, transport, storage, protection, movement and communication. See also, **differentiation** and **bioelectrogenesis**.
EU referendum (2016)	A referendum in which the British people were asked to choose between being '*in, to opt-out*', or being '*out, to opt-in*', as regards **EU** membership (and which resulted in '*no deal*' versus '*no Brexit*').
European Convention on Human Rights	Article 10 provides that "*everyone has the right of free expression*", and that "*this right shall include the freedom to hold opinions and to receive and impart information and ideas, without interference by public authority and regardless of frontiers*". See also, **IMPART**.
European Economic Community (EEC)	The ECSC, EEC and Euratom were established in 1957 (this 'Common Market' laying the foundations for the European Union).
European Union (EU)	The European Commission, based in Brussels, is answerable to every EU citizen: 1) through the European Council (comprising the various heads of government); and 2) through the directly-elected European Parliament. Indeed, the parliament can dissolve the Commission, a Commission proposed by the Council (making the EU democratic, with 'Brussels' its **executive** branch). Its perceived deficiencies reflect voter apathy, media cynicism and a lack of engagement. See also, **Brexit**.
Evidence	See **admissible**.
Evolution Evolutionary Evolutionary laws	Evolution pre-dates **life**. That is to say, evolution 'evolved' the moment that **proteins** and **nucleic acids** attained an equal standing, littering the planet with '**extinct**' organic material. Building upon the same, **macro-molecular machinery** made of proteins, and scripted for by nucleic acids, later succeeded in fabricating the very first cell (whereupon fecundity, coupled with **adaptive radiation**, produced a surfeit of progeny, some possessing advantageous characteristics).

Exceptionality	Denotes a deviation from linguistic or cognitive norms, due to one's abilities or debilities. See also, **colloquialism** and **informality**.
Exchange rate (economics)	The price at which another currency can be bought. Lowering the **interest rate** decreases the exchange rate, i.e. the currency depreciates (negatively affecting imports, but boosting exports).
Executive Executive branch of government	The executive branch of government is centred on the President in American politics, the Prime Minister (and their cabinet) in British politics, and the President of the Commission in European politics. The principal function of the executive is policy-making.
Exergonic (energy-producing)	Molecules which are synthesized spontaneously, e.g. water, are the product of exergonic reactions. See also, **endergonic**.
Existentialism	A philosophy which stresses individual existence, personal autonomy and **free will**. See also, **humanist**.
Exogenous factors	Those factors arising beyond the individual (particularly those with the propensity to corrupt, such as **telepathically-induced effects**).
Exogeny	Originating outside society. See also, **endogeny**.
Experimental testing	The **scientific method** demands that **working hypotheses** are tested experimentally. See also, **peer review**.
Extensive power	Denotes having **power** over many people.
Extinct Extinction	Extinction, like **evolution**, predates life on earth (each initial abortive synergy between **nucleic acids** and **proteins** helping seed life's later genesis). See also, **constructive theory of life's origins**.
Extra-nuclear inheritance	Extra-nuclear inheritance pertains to **mitochondria** and **chloroplasts**, which have their own **DNA** and replicate independently.
Extrasensory (radiofrequency effects)	The sending and receiving of signals by the **cognitive system** (making it a sixth **modality**, but extrasensory), e.g. **remote manipulation**, **mental telepathy** and the **superimposition of persons**. Extrasensory phenomena blur the line separating **rationalism** and **empiricism**.
Extraterritoriality (law)	Arises when approved persons, places or protocols exist beyond **mainstream** law. Extraterritoriality, in the context of this book, effectively distances *Homo sapiens* from the engine of global governance. See also, **NEURON**.
Extreme sceptic	Extreme sceptics view the truth as unattainable, values as objectively baseless and **life** as without purpose. See also, **nihilism**.
Exuberant synaptogenesis	Abrupt neonate neurological development, involving the formation of synapses between **neurons**. See also, **sensitive period**.

F

False consciousness (politics)	A spurious picture of society and the issues affecting the same, falsely presented by those in power, ostensibly to maintain the class structure or prevailing **adjectival social stratification**.
Far-Right Fascism Fascist	Fascism is a political ideology on the extreme **Right** of politics, being blatantly nationalistic, militarily belligerent and habitually racist. See also, **prestige model of extremism**.

213

Federalism Federal unions	Free-market **capitalism** stresses that the **state** should be independent of **economic** production and wealth creation (leading to federal unions and **globalization**, as the state gives way to capitalist logic). Thus, federal unions, e.g. the **United Kingdom**, **European Union** and **United States of America**, which are bound by the free movement of goods, services and people (or goods, capital and labour) remain the *de facto* **sociopolitical superstructures** of our age (independent states having supplanted militias as today's political pawns). Therefore, whether **Brexiteers** admit it, or not, federalism remains an important post-colonial trend. Indeed, the UK's survival depends upon it advancing federalist principles and breaking new ground in respect of **devolution** (even though English **nationalists** would much prefer to concentrate **power**).
Female (biology)	See **feminine**.
Feminine Femininity	American psychoanalyst Dr Robert Stoller (1924-91) argued that the term 'female' pertains to **sex** (whereas the expression 'feminine' pertains to **gender**). See also, **masculine**.
'fight or flight' **response**	See **sympatho-adrenal stress response**.
Filial ramifications	The **genotypic** and **phenotypic** ramifications of cross-breeding.
First law of **thermodynamics**	**Energy** can neither be created nor destroyed (otherwise known as the law of **conservation of energy**).
First measure of **soundness**	Denotes one's commitment to determining the facts, e.g. **sound deduction**, **scientific method** and **hypothetico-deductive cycle**. See also, **second measure of soundness**, **third measure of soundness** and **fourth measure of soundness**.
Fiscal policies (economics)	**Economic interventions** related to **tax revenue** and **government expenditure**. See also, **monetary policies**.
Fission (dark sciences)	The **gravitational force** arises due to the mutual attraction of **nucleons**, such that 'splitting the atom', i.e. fission, causes that attraction to be outwardly expressed. See also, **spectrum of attraction**, **weak nuclear force** and **fusion**.
Flexible immigration	**Economic** reasoning demands flexible immigration, in support of temporary **stabilization measures** (that is, **fiscal policies** and **monetary policies**). Precisely how flexible, however, is today's burning political question? See also, **Brexit**.
Forebrain	See **telencephalon** and **diencephalon**.
Formalized logic Formal logic (analytic philosophy)	Whereby a language is afforded the symbolism of mathematics. Several reasons present themselves: 1) students gain from deconstructing statements; 2) fragmenting sentences allows their arguments to be analyzed numerically; 3) **sound-minded** individuals are able to integrate language, logic and maths; and 4) scientists can **reverse engineer** a given expression, i.e. determine how it arose within the **brain**. See also, **multivibrational matrix**.
Fourth law of **thermodynamics** ('4ᵗʰ Law')	*"The **electric** and **magnetic** fields, acting upon charged particles which are in motion, must be of a prescribed intensity, otherwise the said particles will jump or fall into different **orbitals**, or else realign in the case of atomic **nuclei**"*. See also, **conservation of electronics**.
Fourth measure of **soundness**	Denotes one's ability to correctly interpret the facts, which may mean persisting with doubt. See also, **first measure of soundness**, **second measure of soundness** and **third measure of soundness**.

Free will	The ability to act and make choices, free of compulsion, coercion and harm. A person exercising free will is said to be acting autonomously.
Frequency Frequency modulated (FM)	Frequency is the number of complete waves emitted by a source per second. The human **mind** is frequency modulated, as evidenced by **brain waves**. See also, **unstable multivibrator**.
Frontal lobe/s (anatomy)	The frontal lobes are the seat of **consciousness** (particularly the **prefrontal cortex**, situated behind the brow ridges).
Function (mathematics)	A function denotes the trajectory of a point, when plotted on an **x-y coordinate system**. See also, **dependent variable** and **independent variable**.
Functional groups	Carbon can form into chains, rings and branches (organic structures which then support functional groups comprising other **elements**). It's the size of the carbon structure, relative to the functional groups' properties, which determines the molecule's biological role.
Functionalism (sociology)	A **macrosociological** interpretation of society, which sees the citizenry as a product of the state's **sociopolitical superstructure**. See also, **false consciousness**.
Functionally-nucleated family	Sociologist Talcott Parsons (1902-79) argued that the nuclear family's significance lies in its functional compatibility with highly-developed economies. One shouldn't allow *Homo sapiens'* shortcomings to diminish our appreciation of **social structures**, including the family. See also, **socialized family**.
Functioning proteins (biology)	Biological health rests upon the ability of **nucleic acids** to code for functioning proteins (such **proteins** aiding support, transport, storage, protection, movement and communication).
Fundamental attribution error/s	Occurs when too much blame is attached to an individual's **disposition**, and too little importance given to the demands of their given situation. See also, **telepathically-induced effects**.
Furbishes	The term 'contemporize' is synonymous with furbish, i.e. 'to furbish' is to make things more contemporary. See also, **furnishes**.
Furnishes	To supply or equip. See also, **furbishes**.
Fusion (physics)	The squeezing together of lighter **elements** into heavier ones, due to the action of **gravity** within stars or supernova. See also, **fission**.
G	
Gaia theory James Lovelock (1919–)	An ecological theory which sees the earth, its climate and organisms as part of a synergetic whole (all working together, sufficient to maintain the conditions for life). See also, **competitiveness**.
Galactic Galactic medium Galaxy clusters (dark sciences)	The galactic medium is dominated by **gravity**, and hence the **medium** where **stellar nucleosynthesis** chemically-enriches, manufacturing **elements** of increasing **atomic mass** (either within stars or during supernova events). See also, **black holes**.
Gamete formation Gametes	Gametes are specialist cells which convey **genetic** information from the **male** and **female** during **sexual reproduction**. Each gamete (spermatozoon and ovum, in men and women respectively) contains half the usual number of **chromosomes**, i.e. 23 (but when fused together into a **zygote**, that number rises to 46). See also, **meiosis**.

Ganglia	Clusters of neurons which enhance **radiological**, **electro-chemical** and/or **hormonal** communication. See also, **basal ganglia**.
Gas Gaseousness	The **dark sciences** define **solidity** as the mutual repulsion of **electrons**, making gases, liquids and vapours all technically solids.
Gause's principle (ecology)	An ecological principle which states that no two **species** can simultaneously occupy the same biological **niche**.
Gender Gender convergence Gender dialectic Gender divergence Gender-fabricating variables	Gender is a **structure** (which, at the **microsociological** scale, individuals ought to be at liberty to construct). However, at the **macrosociological** scale, the subject is more contentious. That is to say, *Homo aequipondium*'s gender dialectic diminishes the subject's currency, leading to gender convergence; whereas *Homo sapiens*' unsound **ascription** drives gender divergence. In the final analysis, the question is not whether one is **male** or **female**, or even **masculine** or **feminine**, but rather, whether one is **sound** or **well-balanced**. See also, **hypermasculinity**, **hyperfemininity**, **hypofemininity**, **gender polarization** and '**duality of gender**'
Gender plasticity	The receptiveness of a given individual, through the course of their development, when subject to **gender-fabricating variables**. **Gender plasticity**, like **hominin** plasticity in general, is controversial, due to the fact that much of what is assimilated is at odds with behavioural imperatives. That is to say, as ecological imperatives are mandatory, one's plasticity ought to reflect extrinsic, rather than intrinsic, factors.
Gender polarization Sandra Bem (1944–2014)	Psychologist Sandra Bem (1944-2014) advanced her theory of gender polarization, a thesis which asserts that **socially-constructed** concepts of masculinity and femininity habitually find themselves exaggerated, whereupon gender roles become mutually exclusive. **Logically** propelling men into highly-paid public and political roles, and logically compelling women to accept unpaid family-orientated domestic servitude. Bem was, of course, deconstructing *Homo sapiens*, rather than its replacement. See also, **gender divergence**.
General cognitive pragmatics	Routine thinking differs from professional deliberation, in-so-far as it may contain **irony**, **implicature**, **presupposition** and **metaphor**.
Generalization	Occurs when the meaning of a word or phrase is broadened.
Gene/s Genetic/s Genetic mutations	A gene is a sequence of bases, sited on a strand of **DNA**, which are needed to produce a specific physical characteristic or physiological function. Generally-speaking, a single gene codes for one whole **protein**. Typically, a protein contains twenty different **amino acids**, each coded for by a triplet of bases, or **codon**. Those individual bases come in four varieties, i.e. guanine (G), cytosine (C), adenine (A) and thymine (T). When two complimentary strands of DNA are bound together, into a double-helix, only the following base-pairings are possible, A-T and C-G. Random changes to a gene's sequence of bases may produce **functioning proteins** which are advantageous.
Genotype Genotypic	The **genes** found within an individual organism. See also, **phenotype** and **neurotype**.
Genus (taxonomy)	Closely-related **species**. See also, **binomial nomenclature**.
Germany, Federal Republic of	Established in 1871, this short-lived imperial power's attempted annexation of neighbouring countries ended in **military** defeat. Today, Germany once again blurs its **sovereign** boundaries, but with highly-profitable European-wide cooperation. See also, **Reich** and **European Union**.

G.f (gravity)	Strong **gravitational fields**, occupying the **galactic medium**. See also, **stellar nucleosynthesis** and **black holes**.
g.f (gravitation)	Weak **gravitational fields**, occupying the **intergalactic medium**. See also, **deep-space nucleosynthesis** and **dark matter**.
GIGO (computer science)	'Garbage in/garbage out' (computers being **logical**, rather than fundamentally accurate). See also, **hermeneutics**.
Global financial system (economics)	Revised and updated the global financial system would feedback negatively around political norms and socioeconomic **set-points**. The question is, how can the world economy favour the **balanced-minded** middle-ground? Firstly, **globalization** frustrates the **Right-wing** and **Left-wing**, either by corroding their beloved boundaries or by confounding those who seek to cement them with nationalization. Secondly, the economy exasperates *Homo sapiens*, which is given to **unsound induction**, creating opportunities for *Homo aequipondium*, and its **sound deduction**. Lastly, these economic **clines** and **demes** are **subordinate** and **supraordinate** respectively, rendering the former susceptible to **cruciform coercion**.
Global governance complex (politics)	The collective noun for **supraordinate states**. That is, states which are stratified **neocentrically** for the express purpose of guaranteeing the **balanced-minded** application of **telepathically-induced effects** (effects applied through their **neuro-cognitive** instrument, **NEURON**). See also, **psychodynamical climax**.
Globalization Globalize (economics)	Denotes the emergence of a single integrated global economy, organized around electronic money, satellite communication, 24 hour trading and the internet. Such globalization is the antithesis of European imperialism (whereby **military** adventure amassed **economic** power), in-so-far as economic interdependence renders military power vulnerable to embargos, sanctions and blockades. As such, globalization is the ideal starting point when faced with regional disorder, economic imperatives often transcending ascribed differences.
Government expenditure	See **fiscal policies** and **economic intervention**.
Gravitation (g.f) (dark sciences)	A very weak **gravitational field**, common to the **intergalactic medium** (which inhibits the passage of light, forcing its conversion into **mass-dispersion**). See also, **gravity**.
Gravitational Gravitational field/s Gravitational force/s (dark sciences)	The term gravitational is a generic term, which includes both **gravity** and **gravitation**. Gravitational effects derive from the mutual attraction of **nucleons**, creating a whole **spectrum of attraction** (including the **strong nuclear force**). See also, **protons**, **neutrons** and **inverse square law**.
Gravity (G.f) (dark sciences)	A pronounced **gravitational field**, common to the **galactic medium**, which supports light's propagation (and provides for matter accreting into celestial bodies, stars and **black holes**). See also, **gravitation**.
Ground-state/s (dark sciences)	A **photon** is emitted when an atom's **electric** field weakens, sufficient for an **electron** to fall back into a ground-state **orbital**.
H	
H-bond H-bonding (chemistry)	Polar compounds, like water, have regions of pronounced **positive charge**, while other compounds have regions of pronounced **negative charge**. When these positive and negative regions stick together, Hbonding occurs, enabling substances to dissolve in water.

Healthy mind paradox	The healthy human **mind** has a tendency to deviate from **clinical normality** due to its reliance on **positive-feedback**. See also, **neurotypic sub-classifications**.
Heat death (physics)	See **entropy**.
Heavy metal toxicity (toxicology)	Heavy metal toxicity (involving lead, mercury, manganese, cadmium, etc) is likely to produce a cascade of life-threatening problems, due to its impact on **cholesterol** metabolism.
Hegemonic democracy Hegemony (politics)	Marxist philosopher Antonio Gramsci (1891-1937) viewed the **capitalist** state as a cultural hegemony, and maintained that its class structure owed much to **false consciousness**. Today, the term hegemonic democracy is more apposite, in-so-far as **social action** gifts the **power** to re-structure to a **ruling elite**, ostensibly with the approval of the electorate. However, **national** and **subnational** consensus, in the context of **cruciform calculus**, is all about revealing individual reasoning, primarily through the expedient of **free will**, it's not about surrendering the **sociopolitical superstructure** at the **supranational** scale, however sizable the majority.
Hermeneutical Hermeneutical reasoning	Reasoning involving the formulation of **valid arguments**, **logically** constructed around **unsound** premises. Such reasoning may contain: 1) **illusory correlations** (false relationships); 2) invalid comparisons (bogus parallels); 3) unwarranted leaps (unproven conclusions); 4) latent messages (suggested conclusions); 5) arguing by association (assuming shared attributes); or else 6) lack an obvious conclusion (implicit arguments). Indeed, the hermeneutical often avoid expert opinion, primary source material and **objective** sampling, much preferring consensus over informed disagreement. See also, **hermeneutics**, **epipoliotical reasoning** and **proscriptive reasoning**.
Hermeneutics (scientific field)	The scientific study of **logical minds**, as regards their origin, impact and expression (from *ermineia*, Greek for interpretation).
Heterogametic (X-Y)	Dissimilar **sex chromosomes**, as found in men. See also, **homogametic** (X-X).
Heteronnubial (connubial dualism)	Reciprocal **superimposition**, involving both sexes (a **balanced-minded** female and her male ally might mirror a **socializing** structure and **structurating** agent, with the former enjoying notional seniority, provided they possess an **axial advantage**). See also, **homonnubial**.
Heterozygous (genetics)	If 'T' and 't' are the **genes** for tallness and dwarfism, then an individual with the genes 'Tt' is heterozygous, as regards those **alleles**.
Hindbrain	See **metencephalon** and **myelencephalon**.
Homeostasis Homeostasis hypothesis	A **dynamic equilibrium**, amounting to a **biological singularity**, arguably arose at the conjuncture of emergent hydrologic, carbon, nitrogen and phosphorus cycles (when ambient temperature, atmospheric pressure and chemical molarity were all favourable). Those **equilibrium reactions** later retreating into the cell, where a **plasma membrane**, allied to **primary endosymbiosis**, enabled Eukaryota to self-regulate around life-sustaining physiological **set-points**, using **negative-feedback**.
Hominini Hominin/s	The sub-family *Homininae* contains the taxonomic tribe known as *Hominini* (*Homo* being its sole genus, making both *Homo sapiens* and *Homo aequipondium* hominins).

Homo aequipondium *Homo aequipondium* *amerolipticus* *Homo aequipondium* *epivevatics*	*Homo aequipondium*, as per **trinomial nomenclature**, comprises *Homo aequipondium epivevatics* and *Homo aequipondium amerolipticus* (respectively, **sound-minded** and **balanced-minded** hominins). *Amerolipticus* constitutes a prospective **deme** and *epivevatics* a coevolutionary **cline** (their **supraordinate** status affording them an **absolute advantage**, and with it the prospect of **absolute prosperity**). See also, **aequiponimity**
Homogametic (X-X)	Matching **sex chromosomes**, as found in women. See also, **heterogametic** (X-Y).
Homologous chromosomes	Maternal and paternal **chromosomes**, containing identical numbers of **genes** (governing corresponding characteristics). See also, **meiosis**.
Homonnubial (connubial dualism)	Reciprocal **superimposition**, involving the same **sex**. See also, **heteronnubial**.
Homo sapiens *Homo sapiens* *epipolioticus* *Homo sapiens* *hermeneutics*	*Homo sapiens*, as per **trinomial nomenclature**, comprises *Homo sapiens epipolioticus* and *Homo sapiens hermeneutics* (respectively, **biased-minded** and **logically-minded** hominins). Both *epipolioticus* and *hermeneutics* constitute **clines**, facing evolution *'in the raw'* (within a **subordinate** position, lacking **cruciform coevolution**). Accordingly, they possess only a **comparative advantage** (and achieve, at best, only **relative prosperity**). See also, **sapiency**.
Homozygous (genetics)	If 'T' and 't' are the **genes** for tallness and dwarfism, then individuals with the genes 'TT' and 'tt' are homozygous, as regards those **alleles**.
Horizontal axis (cruciform calculus)	**Cruciform calculus** utilizes the **x-y coordinate system** (it having a horizontal 'x' axis for **sociopolitical superstructure** and a vertical 'y' axis for **neurotype**). See also, **vertical axis**.
Hormonal (endocrine system)	One of the human body's three major **communication systems**. See also, **hormonal imbalance**, **electrochemical** and **radiological**.
Hormonal imbalance Hormonally Hormone/s	The **endocrine system**'s principle organs are the pituitary, thyroid and adrenal glands, together with the pancreas, gonads, hypothalamus, gastro-intestinal tract and pineal gland. Many of the chemical messengers synthesized and secreted by the same are **antagonistic hormones** (with endocrinal disorders giving rise to hormonal imbalance). See also, **electrochemical**, **radiological** and **hypothalamic-pituitary-adrenal axis**.
HPA-axis	See **hypothalamic-pituitary-adrenal axis**.
Human genome (genetics)	Human beings possess 46 **chromosomes**, i.e. 23 pairs of chromosomes (amounting to some 3.2 billion base-pairings, or twenty to twenty-five thousand **genes**). Sequencing of these genes was undertaken by the Human Genome Project (1990–2003).
Humanism Humanist/s Humanist bias	Humanists believe that ethics, science and justice are sufficient to build and sustain a better world. In reality, they aren't, in-so-far as *Homo sapiens* is naturally corrupting of the same. That is to say, ethics, science and justice are **structures**, whereas humans are **agents**. Begging the question, do you believe in the **structures**, or in the agents? Humanists fallaciously claim to have confidence in the agents, whilst secretly admiring the structures (structures subverted by humans). In the eyes of **anti-humanists**, for example, humanism and **religion** both market unrepresentative models, that is, a delusional model of humanity versus a fraudulent depiction of the cosmos. See also, **structuralists** and **Almighty influence**.

Hund's rule (chemistry)	States that every atomic **orbital**, in a given sub-shell, must be partially filled first, prior to the addition of a second **electron**.
Hydrologic cycle	Describes the cyclical passage of water through **phase transitions**, or material states, e.g. ice, fresh water, sea water and vapour clouds.
Hydrophilic	Water-loving (readily forming **H-bonds** with water).
Hydrophobic	Water-hating (avoiding interaction with water, i.e. not **H–bonding**).
Hyperfeminine Hyperfemininity	Extreme **gender**-identification, i.e. exaggerated **femininity**.
Hypermasculine Hypermasculinity	Extreme **gender**-identification, i.e. exaggerated **masculinity**.
Hyperplasia	Increase in the number of cells, i.e. tissue formation.
Hypertrophy	Increase in cell size, i.e. tissue enlargement.
Hypofemininity	Arises when **gender divergence** involves reduced **femininity**, amid **hypermasculinity**, as with the **pecuniary model of extremism**.
Hyponym/s Hyponymic hierarchies	A hierarchy in which **lexemes** (specifically hyponyms), are examples of more general terms, e.g. a man is a hominin, a hominin is a mammal. In a hyponymic sense, a **meme** is a **concept** – or else, **behaviour** we can conceptualize (with **semiotics** the **conceptual framework** providing for their transmission).
Hypothalamic-pituitary-adrenal axis (HPA-axis)	The hypothalamus, sited within the brain's **diencephalon**, integrates the **nervous** and **endocrine** systems, by way of the pituitary gland (making **homeostasis** the HPA-axis' principal function). Using **neurotransmission** and **arteriotransmission**, the hypothalamus instructs the pituitary gland to increase or decrease **hormone** production, thereby regulating thirst, hunger, temperature, alertness, etc. See also, **cognitive-nervous-endocrine axis**.
Hypotheses Hypothesis	Statements which can be tested by experiment and observation, ideally predicting the result. There was a time when many conjectured that "*the sun goes around the earth*", without that hypothesis being strictly refutable – at which point, weak evidence in support of the same would've seen science moving towards an as-yet-unproven nonsense. Engagement with reality is best achieved through refutable hypotheses (which, if not falsified, can be worked-up into theories).
Hypothetico-deductive cycle Hypothetico-deductive research	Hypothetico-deductive research utilizes the **scientific method**, i.e. a refutable **hypothesis** is tested by experiment and observation (the hypothesis often conforming to prevailing theory), leading, in the absence of **empirical** refutation, to a compelling paradigm. The hypothetico-deductive cycle demonstrates that **induction** is best used to generate – not 'theory' – but rather, **working hypotheses**.
I	
Id (Freudian psychology)	The human **brain**'s unconscious source of primitive drives and desires. See also, **ego** and **superego**.
Idealism (philosophy)	Idealism proposes that reality is a mental construct. As the **conceptual** shapes the **corporeal**, mental processes mirror the broader shaping of reality. See also, **materialism** and **conceptual**.

Ideological (power) Ideology	Ideological power hinges upon an electorate's command of **critical thinking skills**, their **neurotypic sub-classifications** and whether they reside in a **disaggregated state**. *Ergo*, a diminution, in respect of all three, greatly increases other people's ideological influence. See also, **political**, **military** and **economic** (power).
Ill-balanced	The ill-balanced "*favour punishing: using reward selfishly, and without obvious limits*" (a position diametrically at odds with **the rule** appropriated from childcare). See also, **pecuniary extremism**, **prestige extremism** and **well-balanced**.
Illusory correlation/s	Algebraic expressions which make **variables** appear instrumental, rather than incidental (the most common being *time* and *distance*).
IMPART (Investigation, Mitigation and Public Awareness of Radiofrequency Technologies)	The writer (mindful of the perils of **remote manipulation**, **mental telepathy** and the **superimposition of persons**) took it upon themselves to commence lobbying on these issues in the 1990s, whilst working as an **Officer of the Supreme Court** (with IMPART, as their campaign became known, leafleting Prime Ministers, MPs and government departments). See also, **WOMEN IMPART**.
Implicature	Expressions with an implied meaning, rather than a literal one.
Inactivation-emission (dark sciences)	The neuron's inactivation-emission phase is **exergonic**, with **energy** lost through **thoughts** and **brain waves** (a process concurrent with **sodium inactivation** and the membrane's repolarization). All active brain cells are conjectured to exhibit **absorption-impulse** and inactivation-emission phases. Making **neurology** inherently frenetic – that is to say, cellular contiguity, chemical suppression and cell death appear essential to normal **brain** function. See also, **unstable multivibrator** and **CN-cycle**.
Inadmissible (law)	Evidence which cannot form part of legal proceedings, e.g. hearsay (in which a person conveys the testimony of another, who isn't available to be cross-examined) is inadmissible. Comments regarding another's thoughts ought to be similarly inadmissible. See also, **admissible**.
Independent **assortment** (genetics)	**Homologous chromosomes**, lying side-by-side in an 'XX' configuration, having exchanged small segments of genetic material via **crossing over**, line-up and separate independently during **meiosis** (producing gametes with **genes** from both parents).
Independent **variable/s** (mathematics)	The variable which alters during experimental testing, enabling one to evaluate its impact on a **dependent variable**. One should avoid **illusory correlations**, as happens when one implies that *time* and *distance* are instrumental, rather than incidental. See also, **Maxwell's equations**.
Induced dipole (physics/chemistry)	Occurs when two molecules are in close proximity, and the **electrons** in one repel the electrons in the other producing a temporary dipole, and thus mild **electrostatic attraction**. See also, **dispersion forces**.
Induction (physics)	See **electromagnetic induction**.
Induction Inductive (logic/philosophy)	With induction observation or **cogitation** comes first, from which 'theory' is then derived; unlike **hypothetico-deductive research**, in which the **hypothesis** comes first, whereupon tests and observations follow. See also, **unsound induction**

Inflation (economics)	Inflation, whereby the price of goods and services increases, is largely due to **demand** exceeding **supply** – as happens when consumption rises (or there's a disruption to the supply chain, as with a 'no deal' **Brexit**). See also, **unemployment**, **stagflation**, and **deflation**.
Informality	Informal deliberation and communication, i.e. **cognition** and self-expression requiring less apprehension as regards grammar, style and content. See also, **exceptionality** and **colloquialism**.
Integral calculus	Type of **calculus** which analyzes displacement, areas and volumes.
Intelligence Intelligent life	Defined by some as an individual's capacity to pro-actively learn and suitably adapt (the antithesis of which is **dumbing down**).
Intensive power	Denotes a high degree of compliance from ones subordinates.
Intercellular connections	The connections between **neurons**: 1) converge (signals pass to fewer neurons); 2) diverge (signals pass to more neurons); and 3) reverberate (signals cyclically persist, as with **memory**). **Neurological development** takes the form of a **mind cycle** (that is, **thoughts** trigger receptive neurons, leading to the excitation of multiplying **neural pathways**, with subsequent **sodium inactivation** radiating further thoughts, and hence spiralling neurological activity, as the potential for **mental illness** grows through time). See also, **unstable multivibrator**.
Interest rate/s	The bank rate (that is, the amount owed to a national bank by a commercial bank, as a percentage of money borrowed) is influenced by the central government's **monetary policies** (which, in turn, affects the interest rates passed on to consumers).
Intergalactic Intergalactic medium	The intergalactic medium is dominated by **gravitation**, and hence the **medium** where **deep-space nucleosynthesis** effectively resets the universe's 'factory settings' (due to **electromagnetic radiation** being reconstituted as the most basic quantum pieces).
Internally valid	See **valid argument**.
Interpretation/s (critical thinking)	Interpretations highlight who is '*less than logical*', '*logical*', '*sound*' or '*more than sound*'. For example, some interpret the attack on Pearl Harbour, in December 1941, as a '*damned good thing*', the Nazis having moved over to mass extermination by gas the previous summer, with continued inaction leading to the liquidation of 10 million Jews and Hitler's possession of the atomic bomb. See also, **fourth measure of soundness, neurotypic sub-classifications** and **perception**.
Invalid argument	An argument is invalid if the conclusion doesn't follow **logically** from the premises (also known as a '*non-sequitur*').
Inverse square law (classical physics)	**Light** intensity is inversely-proportional to the square of the distance from its source, i.e. a doubling of the distance reduces the light's intensity to a quarter of its former value (with **gravitational forces** exhibiting a similar diminution in their strength). As regards 1^+ and 1^- point charges in an **atom**, the **dark sciences** argue that **electrostatic attraction** is inversely-proportional to **energy**, rather than distance. See also, **direct square law** and **Maxwell's equations**.

Ionic Ionic bonding (chemistry)	Occurs when an **atom** donates an **electron** to another atom, creating a strong **electrostatic attraction** between the two (a case of overlapping **magnetic fields** greatly weakening the **electric fields**, causing substantial amounts of **energy** to be released, as **electrostatic attraction** draws the atoms into a lattice).
Iran-Saudi Arabia 'Cold War' (politics/history)	The Middle East is ridden with schisms, many arising from the ongoing Iran-Saudi Arabia 'Cold War'. A 'Cold War' which (like the **Cold War** which previously existed between **The West** and Soviet Union) seeds, aggravates and prolongs civil wars and sectarian struggles, be they Lebanese, Syrian or Yemeni. For example, the Houthi drone attack on Saudi Arabian oil facilities, in 2019, highlights The West's muted loyalty towards Saudi Arabia (and Iran's understated technical support for Yemeni rebels). Of course, Western support for Saudi Arabia will be muted, if it persists in murdering prominent critics.
Irony	Expressions with connotations at odds with a phrase's literal meaning.
Islam (world religion)	Both **Islam** and **capitalism** are pernicious. That is to say, both are **Right-wing** and exploitative in their unchecked forms. However, both have a future within society, provided they're tempered with a **decomposition of power** and a **decomposition of capital**. See also, **Islamic fundamentalism**, **Islamophobia** and **underage brides**.
Islamic fundamentalism Islamofascism	Islamic fundamentalism falls under the **prestige model of extremism**, being on a par with **fascism** in its use of **ascription**, **anti-Semitism** and **scapegoating** (such scapegoating frequently amounting to grotesque public spectacle, often involving the perverse use of amputations). See also, **Islam** and **Islamophobia**.
Islamophobia	Islamophobia parallels **anti-capitalism**, in-so-far as both envisage the demise of exploitative **Right-wing** socioeconomic systems. However, both **Islam** and **capitalism** look set to stay, albeit in a revised form.
Isolated system (physics)	A system which is unable to exchange anything with its surroundings (neither **mass**, **energy**, nor heat). A system, such as the **triune brain** (which is neither an **open system**, **closed system**, nor isolated system), is termed a **semi-isolated system**, in-so-far as it attempts to exchange energy primarily with itself.
Israel	Every Middle Eastern state was established in the 20th century, with Israel no exception (founded in 1948). All **social structures** (e.g. nations, legal systems, etc) are populated by **agents**, many of whom are **unsound**. *Ergo*, irrespective of where one draws the geopolitical lines, one will always end up with intractable problems, due to the overwhelming influence of *Homo sapiens*.
J	
Judiciary Jurisprudence (law)	The principal role of the judiciary is to preside over legal cases arising from alleged transgressions of the law (their competency being directly proportional to their given **neurotypic sub-classification**). Unpardonably, a **sound** verdict, presented to an **ill-balanced** judiciary, results in disproportionate punishment. Therefore, beyond a police force's elementary logic, forensic science's scrutiny of the evidence, the commitment of sound barristers and the open-mindedness of judges, we meet with the need for **balanced-minded** sentencing.

223

K

Keynesian laws Keynesians (Left-wing economics)	Left-wing **Keynesians** believe that **economic growth** follows **logically** from meeting rising **demand**. Whereas, Right-wing **monetarists** caution against encouraging unsustainable demand, given that **inflation** follows. See also, **economic intervention**.

L

Lateralized equity Lateralized extrapolation (cruciform calculus)	**Lateralized equity** arises because deviations from the political **Centre** meet with increasing **economic** vulnerability (a state's **pecuniary** and **prestige** policies assisting with economic extrapolation). See also, **stratified equity** and **economic equilibrium**.
'4th Law' (fourth law of thermodynamics)	The writer's 4th Law states: *"the electric and magnetic fields, acting upon charged particles which are in motion, must be of a prescribed intensity, otherwise the said particles will jump or fall into different orbitals, or else realign in the case of atomic nuclei"*. This **conservation of electronics**, coupled with the laws governing **electrostatics**, guarantees that an atom's **electrons** remain a discrete distance from the **protons** (notwithstanding that if **energy** levels are enormous the electrons must form a **plasma**, or else bind with the protons to produce **neutrons**, if energy is absent).
Laws of thermodynamics (classical physics vs. dark sciences)	It remains an open question as to whether an **isolated system** can confound **entropy**, thereby contradicting the laws of thermodynamics (after all, **deep-space nucleosynthesis, stellar nucleosynthesis** and **black holes** do much to challenge those laws). See also, **first law of thermodynamics, second law of thermodynamics, third law of thermodynamics, fourth law of thermodynamics, 'zeroth' law of thermodynamics** and the **laws of topodynamics**.
Laws of topodynamics (dark sciences)	Like the **laws of thermodynamics**, topodynamics establishes laws regarding **mass-dispersion**, or **space-time**, thus determining if **entropy** is illusory (e.g. Can space be created and destroyed? Does time become unavailable? And is there a minimum space at which time ceases?). See also, $E=mc^2$.
Left Left-wing (political)	**Socialist** ideology, with an emphasis on nationalized industries within otherwise democratic **capitalist** economies. By means of **progressive taxation**, Left-wing politicians seek to boost public finances without imposing upon the poorest in society. See also, **Keynesian laws, Left-wing extremists** and **structuralism**.
Left-wing extremists Left-wing incitement	Left-wing extremists promote **communism** and **Marxist-Leninism**, believing that **capitalism** and **democracy** should give way to one-party dictatorships (any subsequent elections comprising a limited choice of radical Left-wing candidates). Left-wing extremists are predisposed to **situational attribution**, believing that social inequalities justify law-breaking in order to correct systemic injustice. Accordingly, **progressive taxation** morphs **logically** into the liquidation of asset-owning **capitalists**, together with the confiscation of their wealth. See also, **pecuniary model of extremism**.
Lexeme/s	A word, or words, possessing a distinct meaning. Lexemes have a literal meaning, or **denotation**, plus, in many cases, an associative meaning, or **connotation** (which, taken together, encompass all the feelings and ideas attached to that term).

Life Life cycles (evolutionary theory)	Evolution began as a trial-and-error **synergy** between **nucleic acids** and **proteins**. *Ergo*, prior to life on earth there was only evolution and **extinction**, driven by **electromotive** and **nucleomotive** forces, littering the planet with defunct **pre-biotic** material (that synergy later supporting a **dynamic equilibrium** within self-replicating **plasma membranes**). See also, **ontogeny**, **primary endosymbiosis**, and **constructive theory of life's origins**.
Light	See **electromagnetic radiation**.
Limbic system	The limbic system is the principal generator and regulator of **personality** and emotions. See also, **CN-cycle**.
Liquidity Liquid/s	The **dark sciences** define **solidity** as the mutual repulsion of **electrons** (making gases, liquids and vapours all technically solids).
Logic Logical	See **logically**.
Logical connectives (computer science)	AND, OR, XOR, NOT, NAND, etc, are logical connectives (or **logic gates**), whose use of **binary inputs** makes them **Boolean operators**.
Logical constants (linguistic calculus)	Then ($\rightarrow$), not ($\sim$), and (&), or (v), if and only if ($\leftrightarrow$), etc, are logical constants, used to connect propositions in **propositional calculus**.
Logical form	A form of argument which is internally **valid**, and thus arriving at a **logical** conclusion (a conclusion which may not be factually accurate).
Logically Logically-unsound Logical mind/s Logically-minded	Logically-minded individuals conjure-up **valid** arguments (which respect **logical form**), thus arriving at conclusions in-keeping with their unproven premises. Such **unsound induction** is, at best, a source of **working hypotheses**, and therefore excusable within the context of the **hypothetico-deductive cycle**. To put it in perspective, a guilty verdict, following logically from unproven assertions, would be a verdict disowned by **science**. See also, *Homo sapiens hermeneutics*.
Logical operators (mathematics)	Add (+), subtract (-), divide ($\div$), etc, are all logical operators, used to connect **values**. See also, **logicism**.
Logical positivism (philosophy/logic)	Ludwig Wittgenstein (1889-1951) and Bertrand Russell (1872-1970) introduced logical atomism, an approach in which facts represented *'atomic'* building blocks of complex *'molecular'* arguments, *'bonded'* together by sentential operators (later, the **Vienna Circle** consolidated their ideas as **logical positivism**). See also, **verificationism**.
Logic gate/s	See **Boolean operator**.
Logicism (philosophy/maths)	Logicism sees mathematical outputs as **logically** deduced. One can, in fact, produce **biased**, **logical**, **sound** and **balanced** examples of mathematics, mirroring language use (and thus rendering numerical forms of neurotypic analysis easier).
M	
Macro-molecular machinery	The three-dimensional configuration of **proteins** strongly influences their biological roles (a **structure-function** which reflects the size of the carbon framework, **functional group** characteristics, electrical polarity, and **hydrophobic** or **hydrophilic** properties). Macro-molecular machines, made of proteins, machinate under the commanding **conceptual** influence of **electric** and **magnetic** fields.

225

Macrosociological Macrosociology	**Interpretations** which see society as **socially-constructed**. This book's extreme positivism, termed the **ameroliptical perspective**, is a sociological position founded upon the macrosociological premise of increasing **aequiponimity**, amid diminishing **sapiency**, sufficient to make scientific modelling easier through time. That is to say, **ameroliptical** individuals see **cruciform calculus** as coevolving grounds for scientific optimism, with society becoming less resistant to scientific expression. See also, **microsociological**.
Magnetic Magnetic field/s Magnetic field lines (dark sciences)	Circular magnetic field lines, emanating from atomic **nuclei** in close proximity, overlap, strengthening or weakening one another. In a bar magnet that effect is amplified, producing an **electromotive force** on a second bar magnet's electrons. Basically, two bar magnets (with opposite or identical poles meeting) will attract or repel one another, because the **fourth law of thermodynamics** demands that their electrons move closer or further apart. In deep-space, radiating magnetic fields break-down, by virtue of **gravitation**, into **protons** and **gravitational forces** (with **electric fields** producing **electrons** and **dispersional forces**). See also, **configurational theory**.
Magnetic resonance imaging (MRI)	When a person is placed in a strong **magnetic field**, their body's soft tissues emit signals which can be converted into computerized images (that's because the magnet has exerted a **nucleomotive force** on the patient's carbon and hydrogen **nuclei**, bumping their **electrons** up into higher orbitals, with **photons** emitted when they fall back into their **ground-states**). See also, **nucleonic constant**.
Mainstream Mainstream human agency	Mainstream denotes the **subnational** and **national** levels (with the actions of those occupying the same constituting mainstream human agency). See also, **supranational** and **NEURON**.
Male (biology)	See **masculine**.
Market/s (economics)	The **capitalist** market constitutes a **logical** system of commercial exchanges, driven by the compulsion to maximize profits.
Marxism Marxist-Leninism	A form of **pecuniary extremism** (in which a one-party **Left-wing** dictatorship is re-shaped to fit the caprices of its central demagogue).
Masculine Masculinity	American psychoanalyst Dr Robert Stoller (1924-91) argued that the term 'male' pertains to **sex** (whereas, the expression 'masculine' pertains to **gender**). See also, **feminine**.
Mass Mass-dispersion (mc²) (dark sciences)	**Protons** mutually attract creating mass, **electrons** mutually repel creating dispersion (therefore, in order for the universe to expand, **gravity** need only loosen its grip). The **dark sciences** elaborate on the equation $E=mc^2$, postulating that 'E' stands for **energy** and 'mc²' for **mass-dispersion**. That is to say, **energy** exerts an influence via **nucleomotive** and **electromotive** forces, and mass-dispersion via **gravitational** and **dispersional** forces. As **conceptual** and **corporeal** fields comprise isolines and gradients, particles move in all manner of directions – sometimes together, but often apart. See also, **fourth law of thermodynamics** and **magnetic fields**.
Materialism	A philosophy which stresses the importance of the inanimate, from which everything is purported to derive – not least, the animate.
Mathematical pragmatism	See **pragmatism**.

226

Maxwell's equations (classical physics)	Charles-Augustin de Coulomb (1736-1806) formulated **Coulomb's law** in 1785, Michael Faraday (1791-1867) demonstrated **electromagnetic induction** in 1831, and James Clerk Maxwell (1831-1879) published his ground-breaking equations between 1861 and 1873 (showing the relationship between **electricity** and **electromagnetism**). **Differential calculus** illustrates their dilemma, in-so-far as mathematicians can fragment known gradients, whereas scientists must extrapolate from fragmentary experimental data. The **dark sciences** argue that classical physics is relevant, provided one lives on a celestial body, deeply embedded within the **galactic medium**. However, all gradients are deceptive when viewed up-close (and one should avoid attributing to *time* and *distance* the characteristics of **independent variables**).
Medium (physics/chemistry)	The term medium has multiple connotations, e.g. a state between extremes, a material through which energy is conveyed, a conduit for information, etc. See also, **galactic medium** and **intergalactic medium**.
Meiosis Meiotic cell division (genetics)	Whilst **mitosis** is concerned with preserving one's genetic code, meiosis is all about mixing things up during **gamete formation**, by means of **crossing over** and **independent assortment**. See also, **homologous chromosomes**.
Meme/s	The term meme (*mimema*: something imitated) was first coined by evolutionary biologist Richard Dawkins (1941–) to denote an aspect of **semiotics** or **behaviour** which is transmitted in a non-genetic manner. In a hyponymic sense, the **socially-constructed** meme is a **concept**, concepts being aspects of **conceptual reality**.
Memories Memory	See **unistable multivibrator**, **bistable multivibrator** and **multivibrational psychometrics**.
Mental disorder Mental illness (psychology)	Psychiatry (with its emphasis on medication) and psychology (with its commitment to psychotherapy) struggle because the human **mind** derives from conscious, subconscious and unconscious radiofrequency events. Indeed, by the time the **mind cycle** has structured **brain waves** into a juvenile **psyche**, that adolescent persona possesses emergent properties stubbornly resistant to revision. See also, **healthy mind paradox**.
Mental telepathy	The bilateral or multilateral transfer of **thoughts**, by means of **radiofrequency technologies**. See also, **extrasensory**.
Mesencephalon (midbrain)	Controls both the visual and auditory systems, as well as filtering sensory information to the **forebrain**. See also, **NE-cycle**.
messenger RNA (mRNA)	Messenger RNA, produced by **transcription**, contains the **genetic** information pertaining to a single **gene**. See also, **ribosome**.
Metaphor	Figurative phraseology or simile.
Metencephalon (hindbrain)	Associated with posture and motor-learning, and comprising both the pons and cerebellum. See also, **myelencephalon** and **NE-cycle**.
M.f	Strong **magnetic fields**, occupying the **galactic medium**.
m.f	Weak **magnetic fields**, occupying the **intergalactic medium**.
Microorganisms (microbiology)	Unicellular microorganisms (such as nitrifying, denitrifying and nitrogen-fixing **bacteria**) are crucially important to our survival. Such that degrading the soils with pesticides undermines plant growth.

227

Microsociological Microsociology	**Interpretations** which see society as **biologically-pro-duced**. Given its biological importance, the family, whether nucleated or socialized, serves as a microsociological counterpoint to more **macrosociological** constructs – replete with its very own scapegoating, incitement, patriarchy, insurgency and exploitation (making a **decomposition of power** and capital within the family, every bit as important as its broader separation, fragmentation and denial). See also, **humanism**.
Midbrain	See **mesencephalon**.
Militant-Left	See **pecuniary model of extremism**.
Military (power)	Military **power** is wielded coercively against foreign foes and domestic agitators. See also, **ideological**, **political** and **economic** (power).
Mind Mind-body duality	The mind, or **mind cycle**, is a **complex adaptive system**, with **gender** an emergent property and **sex** an initial starting condition. Mind-body duality, or **Cartesian dualism**, reflects the fact that the **corporeal** is subservient to the **conceptual**. See also, **E=mc²**.
Mind cycle (CN-cycle)	When 'fired', a brain cell's **absorption-impulse** phase sends out signals along established **neural pathways** (with **inactivation-emission** causing **thoughts**, or **brain waves**, to be emitted as each **neuron** returns to its resting state). Concurrent with this **positive-feedback** loop, in which **neurotransmission** triggers **radiotransmission**, and *vice versa*, is the laying-down of **memories**, via **neuronal induction**. The mind cycle evolved to shape **placental mammal** behaviour **post-K/T impact**, as evidenced by **Pavlovian conditioning** and **operant conditioning**. See also, **mind** and **consciousness**.
Misandry	Prejudice directed at men. See also, **misogyny**.
Miscegenation	Mixing of the races (or of **neurotypic sub-classifications**).
Misogyny	Prejudice directed at women. See also, **misandry**.
Mitochondria	See **primary endosymbiosis**.
Mitosis Mitotic cell division (genetics)	A human cell contains 46 **chromosomes**, arranged in pairs. When dividing, each chromosome shortens and thickens, revealing itself as 2 **chromatids**, joined at a **centromere** (like a letter 'X'). Consequently there are 92 chromatids – which pull apart, producing two cells, each containing 46 chromatids, whose **genetic** code resembles the original 46 chromosomes. Hey presto, one now has two cells containing identical **genetic** material. See also, **cell cycle** and **meiosis**.
Mitre paradox (law)	The mitre paradox is defined as follows: "*the defence of having been abused is admissible, in the case of an alleged wrongdoing, provided that the abuse can be proven. However, if the abuse is provable, why transgress – why not simply prove that one was being abused?*" This legal paradox presents defendants with a distressing Catch-22 scenario, in-so-far as they may find themselves subject to alarming levels of remote interference, none of which can be proven. Moreover, the law might argue that failure to seek medical advice shows that no distress was present, notwithstanding that a victim of **remotely-induced effects** is more likely to lobby Prime Ministers, MPs and government departments. See also, **IMPART**.
Modality Modalities	A category of sensation related to the five senses, i.e. touch, taste, hearing, smell and vision (**telepathy** is a sixth modality, but **extrasensory**). See also, *modus operandi*.
Modulation	See **frequency modulated**.

modus operandi	A person's manner of working (**empiricists** rely on their **modalities**, whereas **rationalists** rely on their **conceptual processing**). **Neuro-cognitive effects** blur the line, as regards rationalism and empiricism, being both **extrasensory** and **conceptual**).
Molecular bond-state/s Molecules (chemistry)	A number of atoms chemically bonded together (**electronegative** atoms, in close proximity, experience a weakening of their **magnetic fields**, forcing their **valence electrons** into lower, more stable molecular orbitals, with the spontaneous **emission** of **photons**).
Monetarist principles Monetarists Monetary policies Money supply (Right-wing economics)	**Right-wing** monetary policies, involving a national bank's management of **interest rates** and **money supply**, e.g. suppressing **aggregate demand** by means of **austerity**, using **interest rate** rises to encourage savings, reigning-in the **wage-price cycle** by breaking the unions and/or aggressively cutting welfare spending. Left-wing **Keynesians**, in contrast, favour **economic interventions** involving **fiscal policies**. See also, **currency wars**.
Monism Monists	Monism (from the Greek: *monos*, single) is a philosophy which maintains that there's only one kind of substance, fabric or reality.
Morphogeny Morphogenesis	Begins with the act of fertilization, and encompasses the whole of the given embryo or seed's initial development. See also, **ontogeny**.
MRI	see **magnetic resonance imaging**.
mRNA	See **messenger RNA**.
Multi-dimensional deviance	Arises when liberties are taken with, for example, two or more of the following: **semantics**, phonetics, grammar, argument and proof. The **mind cycle**, multi-dimensional deviance and **social learning** often amplify humanity's weakness for **unsound** thinking, thereby undermining otherwise stable **communities**.
Multivibrational matrix Multivibrational psychometrics (neuroscience)	The human **mind** mirrors the **electronic configuration**, its **electric** and **magnetic** fields being similarly pre-eminent. Accordingly, the multivibrational matrix, comprising **thoughts** and **brain waves**, gives rise to **personality** and **consciousness**. By mapping the same, multivibrational psychometrics endeavours to circumnavigate the mind's labyrinthine **neural pathways**, quantized **memory** and parallel processing. See also, **mind cycle**, **frequency modulated**, **configurational theory** and **cognitive-nervous-endocrine axis**.
Mutually dependent (mathematics)	Multivariate analysis, in which A is dependent upon B, and *vice versa*.
Myelencephalon (hindbrain)	The medulla oblongata, or brainstem, which regulates **autonomic** functions. See also, **metencephalon** and **NE-cycle**.
N	
N[+/-]	See **neutron**.
National Nationalism Nationalist/s	The **state**, according to nationalists, should be free of external **power** and control (with extreme nationalists serving that end through belligerence, warring and aggression). See also, **Right-wing**.
Nation state/s	For the purposes of this book, a state or nation state refers to a sovereign state or country, as opposed to a state in the **USA**.

NATO	Formed in 1949, the North Atlantic Treaty Organization presently comprises the **USA**, **UK** and **EU**, together with all their allies (its members and partners being jointly committed to increasing their collective capacity to resist armed attack). See also **Free World**.
Natural selection	Evolution by means of **progressive selection** (but may also denote an absence of evolution, by virtue of **stabilizing selection**). See also, **random selection**.
NE-cycle	See **nervous-endocrine cycle**.
Negative Negative charge/s (dark sciences)	**Deep-space nucleosynthesis** resets the universe's 'factory settings', fabricating negatively-charged **electrons** and positively-charged **protons**, which interact **electrostatically** and **electrically**.
Negative-feedback	Negative-feedback arises whenever conditions oscillate around a **set-point** (every deviation from the norm being systematically countered). **Positive-feedback** being the means by which negative-feedback 'evolves' – that is to say, the **natural selection** or **coevolution** of negative-feedback is what actually shapes *Hominini's* future.
Neocentrically Neocentric state/s Neocentric stratification Neocentrism	Neocentrism is a **sociopolitical superstructure**, sustained by means of a **coevolutionary dynamic**, and which is averse to the **top-down** weakness for despotism and the **bottom-up** predilection for disorder. By stratifying neocentrically (i.e. balanced minds at the top, keenly supported by sound individuals, with logical types graded below them and biased elements at the bottom) a democratic **open system** invites **supraordination**. See also, **cruciform calculus**.
Neo-Darwinian synthesis (evolutionary theory)	Charles Darwin (1809-82) published On the Origin of Species in 1859. Then, in 1866, Gregor Mendel (1822-84) established the principles of inheritance. Followed later by the discovery of **DNA**'s structure, by James Watson (1928-) and Francis Crick (1916-2004). Neo-Darwinian synthesis reflects those later contributions, as regards **evolution**.
Neo-Marxism	Contemporary sociological **interpretations**, which build upon, or otherwise modify, the **structural perspectives** of Karl Marx (1818–83).
Neo-Platonism Neo-Platonist/s (maths/philosophy)	Neo-Platonists view the universe as primarily numerical, implying that most things can be modelled using algebra. See also, **Platonism**.
Nerve cell/s Nerve impulse/s Nervous	See **neurons** and **electrochemical**.
Nervous-endocrine cycle (NE-cycle)	The NE-cycle comprises the antiquated portion of the forebrain (that is, the **basal ganglia** and **diencephalon**), together with the midbrain (**mesencephalon**), hindbrain (**metencephalon** and **myelencephalon**), **autonomic nervous system** and **endocrine glands**. **Electrochemically** and **hormonally** driven, the NE-cycle is dominated by **negative-feedback** (with prospective students advised to build upon the **hypothalamic-pituitary-adrenal axis**, when studying the same). See also, **cognitive-nervous** cycle and **cognitive-nervous-endocrine axis**.
Nervous system/s	See **electrochemical**.
Neural pathway/s	The nature and complexity of neural pathways is more important than the number of **neurons** (with the pathways converging, diverging and reverberating). See also, **cognitive-nervous-endocrine axis**.

Neuro-cognitive (power)	Power which derives from being a custodian of **radiofre-quency technologies**, or from being subject to **telepath-ically-induced effects**. See also, **economic**, **political**, **ideological** and **military** (power).
Neuro-cognitive cybernetics Neuro-cognitive effects	Cybernetics is the technological field concerned with self-reg-ulating forms of **artificial intelligence** (which, in the case of neuro-cognitive cybernetics, implies **cruciform calculus** and its **stabilizing** and **progressive** selection of beneficial **neurotypes**).
Neurodegenerative disease (medical science)	Alzheimer's disease, Huntington's disease, Parkinson's dis-ease and dementia all appear to have sub-cellular origins, implying a biochemical cause, possibly linked to **cholesterol** metabolism and its role in **apoptosis**. See also, **exuberant synaptogenesis**.
Neurogenesis	The growth of specialized **neurons**, from undifferentiated stem cells.
Neuro-linguistic programming	The marrying together of **syntax**, **semantics** and **pragmat-ics**, in a proactive way, so as to shape **cogitation** for the better.
Neurological bit	See **unstable multivibrator**.
Neurological development Neurology	Neurology is that branch of medicine which deals with the **nervous system** (whereas psychology and endocrinology study the accompanying **cognitive** and **endocrine** systems). Generically, neurology encompasses the cognitive, nervous and endocrinal.
Neuromagnetism	Neuromagnetism informs the **brain**'s three-dimensional struc-ture, just as Fleming's left-hand rule explains the motion of charged particles inside an **atom**. See also, **configurational theory**.
NEURON (global politics)	The organization responsible for **neuro-cognitive** power, and hence the sole provider of **telepathically-induced effects** worldwide (English being the official language). See also, **global governance complex**.
Neuronal induction Neuron/s Neurophysiology (dark sciences)	The neurons in **neural pathways** can be triggered in one of two ways: 1) **electrochemically**, via nerve impulses, i.e. **neu-rotransmission**; or 2) **radiologically**, via **thoughts** and **brain waves**, i.e. **radiotransmission**. See also, **unstable multiv-ibrator**, **multivibrational matrix** and **arteriotransmission**.
Neuroplasticities	**Plasticity** reflects one's **neurotypic sub-classification** (**ameroliptical** and **epivevatical** neuroplasticities rating more highly than **hermeneutical** and **epipoliotical** ones; the for-mer's objectivity contrasting with the latter's subjectivity).
Neurotransmission Neurotransmitters	Neurotransmitters are chemical substances which facilitate or inhibit the passage of **nerve impulses** across **synapses** (neurotransmission arising when adjacent **neurons** are activated or triggered). See also, **radiotransmission** and **arteriotransmission**.
Neurotype/s Neurotypic Neurotypic sub-classification/s	Neurotype distinguishes between **species** in terms of their **cognitive** processing of information. Thus, the genus *Homo*, to which the species *Homo sapiens* and *Homo aequipon-dium* both belong, contains untold **phenotypes**, together with several neurotypic sub-classifications. Those sub-clas-sifications (**situational attribution** notwithstanding) reflect: 1) the **soundness** of the **multivibrational matrix**, including its capacity to be **well-balanced**; 2) the **corporeal** structure of the **brain**; and 3) the impact of **hormones** and **autonomic** responses. The sciences of **epipoliotics**, **hermeneutics**, **epivevatics** and **ameroliptics** objectively appraise these factors, in order to understand **biased**, **logical**, **sound** and **balanced** responses. See also, **trinomial nomenclature**.

231

Neutron/s ($N^{+/-}$)	A particle which carries no charge, but possesses the same **mass** as a **proton**. See also, **beta decay** and **dark cycle atomic model**.
News	The news media ought to concentrate on current affairs, leaving past events and unproven aspersions to documentary film makers.
Nietzschean crises Nihilism (philosophy)	Nihilistically-speaking, the truth is unattainable, values objectively baseless and **life** utterly without purpose, if one is governed by **biased** and **logical** instincts (dooming Homo sapiens to Nietzschean crises).
Nitrates Nitrogen cycle Nitrogen-fixation	Nitrogen-fixation makes atmospheric nitrogen available to plants, in the form of nitrates. Consumption of those plants, by animals, enables complex **proteins** to be synthesized (proteins which ultimately decompose, with the help of saprophytic bacteria). Completing the cycle are de-nitrifying **bacteria**, which return nitrogen to the atmosphere.
n.m.f	See **nucleomotive force**.
Non-autosomal	An alternative term for allosomal, i.e. **sex chromosomes**.
Nonlinearities Nonlinearity	See **acute nonlinear event**.
Norm/s	See **set-point**.
Normal curve of distribution	When a **dependent variable** is plotted, biological systems often generate symmetrical bell-shaped curves, aiding prediction.
Normative Normative device	Denotes how things ought to be (a normative device helping to achieve that prospective ideal). However, "*agents subvert structures, unless structures subordinate unsound agents*" (that is, **subordination** aids **competitiveness**).
Nubit	See **unistable multivibrator**.
Nuclear family	See **functionally-nucleated family**.
Nuclei (neurology)	Dense concentrations of **neurons**, all possessing closely-related functions. See also, **basal ganglia** and **diencephalon**.
Nuclei (physics)	See **nucleus**.
Nucleic acid/s	Nucleic acids, such as **DNA** and **RNA**, are polymers (comprising large molecules made up of chemically-bonded smaller molecules). Nucleic acids enable organic **life** to reproduce, grow and develop.
Nucleomotive Nucleomotive force/s (dark sciences)	The nucleomotive force (n.m.f) arises as a consequence of the **fourth law of thermodynamics** (being one of the three forces acting on **protons**, the other two being **gravitational** and **electrostatic**). The presence of **neutrons** affects the way in which those three forces impact **nuclei**, and thus the chemical properties of the **atom**. See also, **periodic table**.
Nucleonic constant/s (dark sciences)	If one imagines the **magnetic** and **gravitational** fields as comprising isolines and gradients, then the nucleonic constant forces **nucleons** to move in response to the same. See also, **electrostatic attraction**.
Nucleon/s	**Protons** (and those **neutrons** bound within atomic **nuclei**).
Nucleus (biology)	That part of the cell which contains the **chromosomes**, and hence the **genetic** code for the individual.
Nucleus (physics)	The minute positively-charged core of an atom, comprising both **protons** and **neutrons**. See also, **nucleomotive force**.

O

Objective Objective truth	Ascertaining how things actually are, however imperfect (the objective truth often repudiating **interpretations** and **hypotheses**).
Oceanic factors	Factors pertaining to the marine environment, which influence the world's oceans and the organisms which inhabit the same.
Officer of the Supreme Court, in England and Wales	A person authorized to administer any oath, or take any affidavit, required for the purpose of proceedings in the Supreme Court, in England and Wales. See also, **IMPART** and **WOMEN IMPART**.
'One true' competitive exclusion	If two closely-related **species** of hominin coexist in a finite space, with limited resources, the least competitive will be eliminated. **Cruciform calculus** preempts that 'one true' **competitive exclusion**, through the expedient of **subordination**. See also, **Almighty influence**.
Ontogeny	Ontogeny pertains to a plant or animal's life-cycle, from the moment of their conception to the instant of their death. See also, **morphogeny**.
Open system/s (chemistry/biology)	An open system is able to exchange heat, **energy** and **mass** with its surroundings. See also, **closed system** (physics/chemistry).
Open system/s (sociology)	An open system supports social mobility, thereby avoiding the inequalities, tensions and repression common to **closed systems**.
Operant conditioning	Building upon **Pavlovian conditioning**, in which **placental mammals** associate ostensibly unrelated stimuli, operant conditioning is concerned exclusively with reward and punishment. However, "*favouring rewarding; using punishment sparingly, and only to define the limits*" reflects largely on the conditioner (that is, one should reward those who reflect this **rule**, however short-lived the conditioning).
Operators	See **logical operators**.
Orbital/s (physics)	Each atomic shell or subshell contains a specific number of orbitals, each one of which is capable of holding just two **electrons**.
Organelle/s (biology)	Sub-cellular bodies which perform a range of specialized functions inside the cell. See also, **mitochondria** and **chloroplasts**.
Organic decomposition (ecology)	Organic decomposition exemplifies **negative-feedback**, with the **elements** which sustain **life** being recycled around **biogeochemical norms** (which contrasts with non-degradable plastics, which amass in the manner of **positive-feedback**). See also, **anthropocene**.
Organic molecules (chemistry)	Molecules containing carbon are instrumental in all biological processes (carbon atoms readily forming into chains, branches and rings). See also, **functional groups**.
Organization (cellular biology)	In addition to **hypertrophy**, **hyperplasia** and **differentiation**, one also finds organization, and hence organs, tissues and systems.
Oscillating (physics)	A periodic displacement around an equilibrium point, such as the displacement of a wave around a flat-calm position.
Overproduction (economic)	*Homo aequipondium* avoids overproduction (unlike *Homo sapiens*, which floods the world **market**, while restricting domestic supply).

Oxidation (chemistry)	Oxidation arises when a compound loses an **electron**, and another gains it through reduction.
Oxygenating **photosynthesis**	Green plants produce sugar and oxygen from carbon dioxide and water, using energy from sunlight. See also, **ATP**.
P	
P⁺	See **proton**.
Paradox Paradoxical	Self-contradictory statements, e.g. the **healthy mind paradox** arises because illness is the **logical** corollary of health, and the **mitre paradox** arises because the grounds for acquittal make transgressing unnecessary.
Parasympathetic (nervous system)	The **autonomic nervous system**'s parasympathetic branch 'inhibits' (whereas the **sympathetic** branch 'excites').
Parsimony	Antonym of **profligacy**.
Particle acceleration Particle physics	Particle physics perceives particles as interacting directly, or through intermediary particles (particle acceleration becoming the vogue from the 1930s onwards, with energies measured in **electron volts**). See also, **corporealism**.
Passion	From the Latin '*passionem*', meaning suffering.
Pathological **asymmetries** Pathological forecasting Pathological stratification	**Negative-feedback** dominates **life** on earth, providing for stable **ecosystems** and physiological health. *Homo sapiens* is a global 'pathogen', due to its **personality** and **behaviour** introducing destabilizing **positive-feedback**, making the **anthropocene** notorious for its pathological asymmetries – not least, pathological stratification (a form of **adjectival social stratification**, in which **biased** and **logical** minds subvert **sound** and **balanced** ones).
Patriarchies (politics/sociology)	Societies in which **economic, political, ideological** and **military** power is appropriated by men, for their own benefit. Patriarchies habitually "*favour punishing; using reward selfishly, and without obvious limits*" – leaving women bereft of these principal sources of **power**. See also, **misogyny**.
Pauli exclusion principle (chemistry)	States that no more than two **electrons** can occupy a given sub-atomic **orbital**.
Pavlovian **conditioning**	The association of ostensibly unrelated stimuli (which contrasts with **operant conditioning**, which uses reward and punishment).
Pecuniary Pecuniary extremism Pecuniary model of extremism	A form of political extremism in which **economic** concepts fuel revolutionary beliefs, with the greatest violence aimed at wealthy land-owning **capitalists** (**Left-wing incitement** to commit such infractions stemming from **situational attribution**, whereby unconscionable inequalities are cited as justifying such violence). Pecuniary extremism appeals to *Homo sapiens*, due to the lure of its **logic**. See also, **Communism, Marxism** and **prestige extremism**.
Peer review	Peers are experts in a particular field, who are asked by editors to review academic papers (with editorial interference, unconscious favouritism and unwarranted trust cited as deficiencies). The only alternative being replication using **double-blind controls** (variously described as replicability, reproducibility and repeatability).
Peer-to-peer **communication**	See **social learning**.
People's Republic of **China** (PRC)	Mao Tse-tung (1893-1976) declared mainland China to be the communist People's Republic of China in 1949, following a bitterly-fought civil war. His nationalist opponents retreated to the island of Taiwan, where they established the Republic of China (ROC).

Perception	Perception combines *bottom-up* **modality**-driven processes, related to the five senses, with *top-down* **conceptual processing**, drawing heavily on experience and learning. See also, **interpretation**.
Periodicity Periodic table	One of the horizontal rows in the periodic table of **elements** (a table which organizes the elements according to their atomic numbers). Comparative periodicity, as regards **electronegativity**, may reflect annular lines of **magnetic** influence in **atoms** of escalating **atomic weight**.
Personality	See **cognitive-nervous cycle** and **emotions**.
ph*/-	See **photon**.
Phases of matter Phase transition/s (physics/chemistry)	A phase transition indicates a material change, with the **conceptual** manipulating pressure, temperature, molarity, **potential difference**, and so on (turning **solids** into **liquids** and liquids into **gas**, etc).
Phenomenologists Phenomenology	Phenomenologists are extreme **anti-positivists**, who see the **science** of society as impossible due to *Homo sapiens'* characteristically subjective **behaviour**, corrosive **social learning** and inadequate **critical thinking skills** (making phenomenology a **microsociological** perspective). See also, **ameroliptical perspective**
Phenotype/s Phenotypic	The manner in which an organism's **genotype** is physically expressed (as opposed to the **genes** it possesses). See also, **neurotype**.
Phosphorus cycle	Phosphorous, an essential component of **nucleic acids**, is continually recycled by means of **organic decomposition** and the weathering of sedimentary rocks.
Photon/s (ph*/-)	An elementary quantum of action (**electromagnetic radiation** possessing the characteristics of both a particle and a wave).
Phyla Phylogeny	Phyla, plural of phylum, is a taxonomic expression, falling between kingdom and class (the academic study of those progressively radiating **evolutionary** relationships being termed phylogeny).
Physics	Orthodox physics focusses on the forces which maintain and conserve the **electronic configuration**. See also, **chemistry**.
Placental mammal/s	Placental mammals give birth to fully-formed live young. More significantly, one can think of them as evolving a **triune brain**. The triune brain – an idea introduced by American neuroscientist Paul MacLean (1913-2007) – interprets the mammalian **brain**, and hence those of humans, as comprising a reptilian complex, paleomammalian complex and a neomammalian complex. That conjuncture of the **limbic system** (paleomammalian complex) and **basal ganglia** (reptilian complex) being pivotal, as regards the proposed **CNE-axis**.
Planck's constant (*h*)	Has the value 6.626 x 10^{-34} js (thus, the **energy** in a quantum of light is given by $E = hv$, where *v* is the frequency).
Plasma (dark sciences)	The **fourth law of thermodynamics**, plus ambient energy and the principles governing **electrostatics**, guarantee that **electrons** remain a discrete distance from atomic **nuclei** (such that if **energy** levels are enormous the electrons will form a superheated plasma).
Plasma membrane/s (biology)	The semi-permeable membrane enclosing a cell (sufficient to contain metabolic processes and **equilibrium reactions**). See also, **cell wall**.
Plasticity	The malleability of a person's character, throughout their development, when subject to inculcation, experience and stress. See also, **neuroplasticities**, **gender plasticity**, **placental mammals**, **operant conditioning** and **CNE-axis**.

235

Platonism (maths/philosophy)	Plato's (c429-347BC) view that mathematical objects are objectively real, irrespective of humankind's presence. See also, **neo-Platonism**.
Plebiscite/s	The canvassing of an electorate's opinion on a specific issue.
Pluralist 'tabula rasa'	The democratic 'blank slate' (which, from its inception, lends itself to **cruciform calculus**, and hence to the **coevolution** of **biologically-produced** and **socially-constructed** factors).
Polemic Polemicist	A strongly-worded argument, which avoids, as far as possible, putatively prefixing statements, in the interests of style, flow and readability. A polemicist's motives may be dispositional or situational.
Political (power)	Elective dictatorship, whereby the ruling party dominates the legislative body (leaving the constitutional structure vulnerable to **political bias** and **political logic**, and hence to **pecuniary** and **prestige** extremism), is the acme of political **power** in a **democracy**. See also, **ideological**, **military** and **economic** (power).
Political bias	**Invalid arguments**, constructed for political effect, which have conclusions at odds with their premises. See also, **political logic**.
Political Centre	See **Centre**.
Political logic	**Valid arguments**, constructed for political effect, which rely upon **ideological** presuppositions. See also, **political bias**.
Political meridian Political middle-ground	The political middle-ground is a **socioeconomic** environment first, a **political** system second and a **military** power third (making it a political meridian from the perspective of **cruciform calculus**, as it comprises a superabundance of **sound** and **balanced** exchanges,).
Political perspective	Whether a **microsociological** or **macrosociological** interpretation of society is pertinent really depends upon the way in which the political perspective informs the **sociological perspective**, and *vice versa*.
Positive Positive charge/s (dark sciences)	**Deep-space nucleosynthesis** resets the universe's 'factory settings', fabricating positively-charged **protons** and negatively-charged **electrons**, which interact **electrostatically** and **electrically**.
Positive-feedback	Positive-feedback arises whenever conditions deviate from a **set-point**, resulting in a further deviation from the optimal (indicating, in the case of human physiology, the need for urgent medical attention).
Positivists	Positivists, like **structuralists**, see society as being a product of its permanent **structures**, **economic** systems and data acquisition, making predictive modelling conceivable. See also, **ameroliptical perspective**, **anti-positivists** and **phenomenologists**.
Post-K/T impact	An asteroid impact, 65 million years ago, fundamentally altered the global climate (triggering, it is believed, the **extinction** of the dinosaurs and the **evolution** of the mammalian **triune brain**).
Post-radical feminism	Women are more **aequipine** than men, on average, thanks to some long-standing **social structures**, e.g. the **functionally-nucleated family**. Post-radical feminism acknowledges that the best of women's instincts derive from trying circumstances, transcending liberation.
Potential difference (electronics)	**Electric fields** induce electrical currents, due to potential differences arising in those fields (differences which mirror subatomic variations, as per **configurational theory**).

Power (politics/sociology)	Defined by Max Weber (1864-1920) as the ability to impose one's will, even when opposed. Implying that power (be it **economic**, **political**, **ideological** or **military**) involves some degree of coercion.
Pragmatics (semiotics)	How a person employs signs, symbols and words when communicating. See also, **cognitive pragmatics**.
Pragmatism (mathematics)	Science's reliance upon probability, extrapolation and statistics gives credence to **conceptualism**'s view that objective reality defies algebraic expression, hence the need for mathematical pragmatism. See also, **abstraction**.
Pre-biotic	See **constructive theory of life's origins**.
Pre-Cambrian	Close on 90% of the earth's history (prior to the **Cambrian explosion**). As all sub-cellular biochemical reactions are **equilibrium reactions**, perhaps a single point on the earth's pre-Cambrian surface substituted for the **cell** (those reactions then retreating into the **open system** we term the cell, where **dynamic equilibrium** is more easily managed).
Predicate calculus	An example of **formalized logic**, which uses letters (e.g. x, y, z) to denote **variables** and quantifiers to denote amounts. See also, **logical atomism**.
Predict Prediction/s	The 'holy grail' of anyone seeking to replace **hypothetico-deductive research** and evidence-based analysis. See also, **retrodiction**.
Prefrontal cortex (neurology)	Anterior portion of the frontal lobe, just behind the brow ridge (where one's **superego**, **personality** and **consciousness** are centred).
Pregalactic medium (dark sciences)	**Deep-space nucleosynthesis** arises in the **intergalactic medium**, and **stellar nucleosynthesis** in the **galactic medium** (making a pregalactic medium appear improbable, given that dichotomy).
Pregnenolone ('master' steroid hormone)	The human body uses **cholesterol** to synthesize pregnenolone, from which glucorticoids, mineralcorticoids and the sex hormones (progesterone, testosterone and oestrogen) are produced. These **antagonistic hormones** regulate, amongst other things, secondary sexual characteristics, pregnancy, glucose levels, mineral balance, electrolyte concentrations and anti-inflammatory responses.
Prescriptive Prescriptive reasoning Prescriptivism	**Critical thinking skills** which ought to be supported and encouraged, i.e. **epivevatical reasoning** and **ameroliptical reasoning**.
Prestige Prestige extremism Prestige model of extremism	A form of political extremism in which fanaticism is driven by perceived differences in social standing, the greatest violence being reserved for those alienated on the grounds of ethnicity, religion and political conviction (with **Right-wing scapegoating** stemming from **dispositional attribution** and the habitual use of **ascription**). Prestige extremism appeals to *Homo sapiens*, due to the lure of its **logic**. See also, **fascism**, **Islamofascism** and **pecuniary extremism**.
Presupposition/s	Key assumptions underlying a person's arguments.
Primary endosymbiosis (evolutionary biology)	The theory that some early **prokaryotic cells** were specialist in **oxygenating photosynthesis** and others in the **oxidation** of glucose, prior to both becoming **organelles** (termed **chloroplasts** and **mitochondria**) within much larger **eukaryotic cells**.
Primary labour market (economics)	Includes, amongst other occupations, scientists, engineers and civil servants (who enjoy job security, healthy remuneration and career prospects). See also, **secondary labour market**.

Primary responsibility	The primary responsibility of **well-balanced** individuals is to political freedom, within an **open system** possessing social mobility.
Procedural knowledge	Established methods, including forms of logic, algebraic formulas and industrial processes. See also, **declarative knowledge**.
Product/s (chemistry)	Substance arising when **reactants** take part in a chemical reaction.
Profligacy	Antonym of **parsimony**.
Programming logic	The trailer of a piece of source code is executed whenever certain conditions are met, making the result a **logical consequence** of the same. See also, **hermeneutics** and **validity**.
Progressive Progressive fecundity Progressive selection	**Evolution** occurs when **stabilizing selection** morphs into progressive selection, and an ecological **community** begins to perceptibly alter. Progressive fecundity, which discriminates against the established **neurotypes**, is the dominant feature of a community which is perceptibly **coevolving**. See also, **clines** and **random selection**.
Progressive taxation (Left-wing economics)	Arises when the **Left-wing** taxes the **primary labour market**, thereby boosting public finances without imposing upon the poorest.
Prokaryotes Prokaryotic cells	Single-celled **archaea** and **bacteria**, which do not possess a discrete **nucleus,** instead their **genetic** material resides in the **cytoplasm**.
proletarians Proletariat	Refers to the working class (which is forced to sell its labour to meet the basic requirements of life, without the security of capital).
Propositional calculus Propositional symbols Propositions	Propositional calculus uses **logical constants**, e.g. then ($\rightarrow$), not ($\sim$), and (&), or (v), if and only if ($\leftrightarrow$); propositional symbols, e.g. p, q, r (in lieu of propositions); and brackets (to aid clarity). Making expressing **logical for**m and deconstructing arguments more efficient.
Proscriptive Proscriptive reasoning Proscriptivism	**Critical thinking skills** which ought to be avoided or discouraged, i.e. **epipoliotical reasoning** and **hermeneutical reasoning**.
Protein/s Protein synthesis	Proteins (which are essential components of living organisms) are nitrogenous compounds. A **gene** is copied from a strand of **DNA** by **transcription**, producing **mRNA**. Then, a portion of that mRNA (or gene), termed a **cistron**, is copied by a ribosome, in a process termed **translation**, producing a protein. See also, **synergy** and **nitrogen cycle**.
Proton/s (P^+)	A particle which carries a single **positive charge** (and the same **mass** as a **neutron**). Together, protons and neutrons form the **nuclei** of atoms. See also, **gravitational force**.
Psyche (Freudian psychology)	The psyche, or self, is said to comprise an **id** (**behaviour**), an **ego** (**consciousness**, **memory** and **emotions**), and a **superego** (**personality**). Which parallels the **triune brain**, comprising a reptilian complex, a paleomammalian complex (or **limbic system**) and a neomammalian complex. See also, **cognitive-nervous-endocrine axis**.
Psychodynamical climax Psychodynamical momentum	A **global governance complex** exemplifies psychodynamical climax, being comparable to a **climax community**, subject to **stabilizing selection** (save-and except that it utilizes **stabilizing fecundity** to maintain **aequiponimity**). Psychodynamical momentum, whereby a 'blip' is seen to migrate in **cruciform calculus**, assists with sociological **prediction**. See also, **psychostationary standing**.

Psychokinesis	That aspect of **superimposition** which is concerned solely with movement (in sport it could be performance enhancing).
Psychostationary standing	**Cruciform calculus** has its very own **'uncertainty principle'**, i.e. one can establish **psychostationary standing** or **psychodynamical momentum**, but not ascertain both simultaneously (psychostationary standing denoting the position of an otherwise migrating 'blip').
Pure-breeding	Pure-breeding *Homo sapiens* comprises 'x-x' and 'x-Y', whilst pure-breeding *Homo aequipondium* comprises 'X-X' and 'X-y'.
Q	
QED	See **quantum electrodynamics**.
QSD	See **quantum spatialdynamics**.
Quantum electro-dynamics (dark sciences)	Quantum electrodynamics (QED) studies **electrons**, **electric fields**, **emission** and **dispersional forces**. **Electromagnetism** and **spatialgravitism** (that is, **induction** and **electrostatics**) constitute the interface between **QSD** and QED. See also, **electricity**, **energy** and **deep-space nucleosynthesis**.
Quantum spatial-dynamics (dark sciences)	Quantum spatialdynamics (QSD) studies **nucleons**, **magnetic fields**, **absorption** and **gravitational forces**. **Electromagnetism** and **spatialgravitism** (that is, **induction** and **electrostatics**) constitute the interface between QSD and **QED**. See also, **nucleus**, **mass** and **stellar nucleosynthesis**.
R	
Race/s Racism	Race denotes a person's **phenotype** (there being no objective basis for racially classifying people on the grounds of **genotype**). However, racists are arguably identifiable on the basis of their genotypes, with no single phenotype responsible for **unsound** racial stereotyping. See also, **demes**, **clines** and **CNE-axis**.
Radicalization (sociology)	Radicalization represents a failure of **acculturation** or **trans-culturation** (with home-grown extremism, in a **hegemonic democracy**, arising out of a failure of **socialization**).
Radiofrequency effects Radiofrequency technologies	Radiofrequency technologies exploit the human **brain**'s latent telepathic potential (which arises from its person-specific **absorption profile**). The range of effects which are possible includes **remote manipulation, mental telepathy, superimposition of persons, psychokinesis** and **telekinesis**.
Radiological (cognitive system)	One of the human body's three major **communication systems** (as each **neural pathway** is laid-down the **brain** is showered with radio waves, triggering further signals, and eventually **consciousness**). See also, **radiologically, hormonal** and **electrochemical**.
Radiologically Radiotransmission	Radiofrequency wavelengths and extra-low frequencies, in the form of **thoughts** and **brain waves**, drive the **cognitive system**. In that sense, negligible **energy** has a sizable material effect, through the **anthropogenic factors** it gives rise to. See also, **radiological, neurotransmission** and **arteriotransmission**.
Random selection	Random selection denotes an absence of **natural selection** (natural selection being either **progressive** or **stabilizing**).

Rationalism Rationalist/s	Rationalism sees knowledge as accruing from reason (its *a priori* approach harnessing language, mathematics and logic to explore and understand the truth). A person who affords **efferent signals** more weight and importance than **afferent signals** is rationalist by nature. See also, **empiricism** and *modus operandi*.
Reactant/s Reactivity (chemistry)	**Products** arise when reactants take part in spontaneous **exergonic** reactions, releasing **energy** in the process.
Real electron flow (electronics)	The path taken by **electrons** in an **electronic** circuit, i.e. from the **negative** to **positive** terminal. See also, **conventional current flow**.
Realism Realist/s (maths)	The view that mathematical objects are real, however innumerate the analysis. See also, **anti-realism** and **logicism**.
Received learning	The antithesis of **social learning**, involving the use of **aequipine** standard texts, i.e. ones which are sound and well-balanced.
Recessive (genetics)	A recessive **gene** can only express itself in its **homozygous** form (a form termed a **double recessive**).
Regressive taxation (Right-wing economics)	**Right-wing** politicians, keen to cultivate support from bourgeois factions, may introduce taxes which fall disproportionately upon the poor, i.e. **secondary labour market**.
Reich (politics/history)	Literally translates as 'Germanic realm', hence this book defining Reich in unbroken chronological terms, i.e. First Reich (prehistory-1871), Second Reich (1871-1918), Third Reich (1918-1945), etc. See also, **Germany**.
Relative poverty (economics/sociology)	In affluent communities some experience relative poverty, e.g. lack of broadband connectivity. See also, **absolute poverty**.
Relative prosperity (economics/sociology)	Relative prosperity denotes an accumulation of capital within an otherwise unsustainable **subordinate** position, be it **pecuniary** or **prestige** in nature. See also, **absolute prosperity**.
Religion	Religion, like atheism, attracts **biased** and **logical** minds, with both atheists and theists harbouring delusions. See also, **humanism**.
Remotely-induced effects	See **telepathically-induced effects**.
Remote manipulation	The direct manipulation of a person, by way of **radiofrequency technologies**, sufficient to alter their **radiotransmission**, **neurotransmission** and **arteriotransmission**.
Reproductive health (medicine)	Heavily regulated by **antagonistic hormones**, reproductive health can be undermined by defective **cholesterol homeostasis**.
Retaliatory logic	Vengeance, retribution and scapegoating exemplify retaliatory **logic**. But none of these **unsound** responses is sufficient to offset the cost of **subordination**.
Retrodict Retrodiction/s	Imagine reverse engineering an **electronic** circuit, together with its inputs, using only the observed outputs (making **hypothetico-deductive research** and evidence-based analysis redundant). One would have to eliminate all implausible events, or **logic gates**, together with every implausible output. See also, **prediction**.
Ribosome/s (molecular biology)	A **cistron** is copied by an organelle called a ribosome in process known as translation, producing a **protein** (that is, the ribosome uses the cistron to bring together different tRNA molecules, and their assorted **amino acid** appendages, to produce a functioning protein).

240

Right Right-wing (political)	A **capitalist** ideology which promotes the capital-owning upper-middle classes, patriotism and **military** strength. By means of **regressive taxation**, Right-wing politicians seek to boost public finances without alienating the bourgeoisie. See also, **monetarist principles**, **Right-wing extremists** and **structuralism**.
Right-wing extremists Right-wing scapegoating	Right-wingers fail to grasp that **capitalism** is fraught with difficulties, such as long-term **unemployment**, which advocates of the **markets** really ought to sympathize with. Put bluntly, "*promote smoking in a confined space if you must, but kindly sympathize with the victims of passive smoking!*" Analogies aside, Right-wing extremists, with their penchant for **dispositional attribution**, are prone to **fundamental attribution errors**, **actor-observer bias** and **denunciation**. See also, **systemic bias** and **prestige model of extremism**.
RNA (genetics)	See **messenger RNA**.
Rule, the	Favour rewarding; use punishment sparingly, and only to define the limits. This rule comes originally from childcare, placing nurture at the heart of this book's professed global values. See also, **well-balanced** and **ameroliptics**.
Ruling class ideology Ruling elite	Multi-party systems oscillate around a political **norm**, due to the impact of **anti-exploitation bias** and **anti-insurgency bias** (which, together, denote the ideological **Left** and **Right** respectively).
Russia Russian Federation Russian/s	Since 1991, 'Russia' has been synonymous with the Russian Federation (the term 'Russians' applying to those wielding **power** in Moscow or to their agents in lands falling under their direct control). See also **Russo-Shia bloc**.
Russo-Shia bloc Russo-Shia bloc subordination	Comprises the Russian Federation, Syria's Assad regime, Lebanon's Hezbollah and the Islamic Republic of Iran (any or all of which may meet with **subordination**). See also, **Americano-Sunni bloc**.
S	
Sapiency Sapient Sapient economics	*Homo sapiens*, and hence sapiency, has given us, amongst other things, transatlantic slavery, weapons proliferation, targeted genocide, mass extinctions and climatological change – the truth being unattainable, absolute prosperity unachievable and competitiveness unreachable if one is governed by **biased** and **logical** analysis. See also, **nihilism**, **comparative advantage** and **aequiponimity**.
Sceptic Sceptical	A person who believes the truth is unknowable, due to deficient abstract models, the illusory nature of reality or simply strategic deception. See also, '*cogito, ergo sum*'.
Schizoid Schizophrenic	A major psychiatric disorder, characterized by a loss of contact with reality, auditory hallucinations and social withdrawal. See also, **fundamental attribution error**.
Science Scientific method Scientific mind Scientists	A **working hypothesis** is tested by experiment – if falsified, a new hypothesis called-for, otherwise it can be refined into a theory. For example, the **hypothesis** that "*tomorrow the sun will rise*" can be worked-up into a theory that "*every day the sun will rise*" (illustrating that theories shouldn't be taken for granted in perpetuity). Hence, the scientific mind, which constitutes **sound-mindedness** at its most penetrating, mirrors the **hypothetico-deductive cycle**. See also, **Maxwell's equations**.
Secondary labour market (economics)	Predominantly unskilled work (ridden with job insecurity, low wages and limited prospects). See also, **regressive taxation**.

Secondary responsibility	The secondary responsibility of **well-balanced** individuals is to the balanced **decomposition of power**.
Second law of thermodynamics	The second law of thermodynamics states that the **entropy** of an **isolated system** increases over time (implying that the universe will one day become inert, with a paucity of available **energy**). See also, **laws of topodynamics**.
Second measure of soundness	Denotes a person's ability to handle a deficit of information, including **sensory deprivation** (as exemplified by **Abelard's** dictum and '*cogito, ergo sum*'). See also, **first measure of soundness**, **third measure of soundness** and **fourth measure of soundness**.
Security Council (UN)	Comprising permanent and non-permanent members, the Security Council is the principal instrument of the **United Nations** (an instrument tasked with maintaining international peace and security).
Semantics (semiotics)	The precise denotations, definitions and meanings of signs, symbols, and words.
Semiconservative replication (genetics)	The mechanism by which **DNA** replicates itself. Provided free nucleotides are available, together with the requisite **enzyme** (DNA polymerase), **energy** (**ATP**), and a moderate temperature, DNA's two helical strands can 'unzip', allowing free nucleotides to attach themselves, creating two new double-helixes. See also, **chromatids**.
Semi-isolated system (physics)	A system which is neither open, closed, nor isolated. The human **brain** attempts to exchange little or no **energy** with its surroundings, only **mass** (though it will absorb and emit energy within itself).
Semiotics	The **conceptual** cornerstone of the **CNE-axis**, comprising **syntax**, **semantics** and **pragmatics**. See also, **formalized logic**.
Sensitive period (child development)	The optimal period for language development is from birth to 6 years of age (particularly 1-3 years, due to **exuberant synaptogenesis**). Regrettably, it can't be pharmaceutically prolonged!
Sensory awareness Sensory deprivation	Multi-sensory impairment cruelly confounds **semiotics**. But even when one's **modalities** assimilate **syntax**, **semantics** and **pragmatics**, we remain susceptible to a scarcity of information. Hence the need for **critical thinking skills**.
Set-point/s	Living systems harness **energy**, in the form of **ATP**, in order to achieve **norms** at odds with concentration gradients. See also, **homeostasis**.
Sex (biological) Sex chromosomes (genetics)	The average human cell has 23 pairs of **chromosomes** (one pair are the allosomal sex chromosomes, the other twenty-two **autosomal** pairs being common to both sexes). In humans, the sex chromosomes are labeled 'X' and 'Y', giving rise to **heterogametic** males (X-Y) and **homogametic** females (X-X). A person's **sex** is **biologically-produced**, whilst their **gender** is **socially-constructed**.
Sexism	Takes the form of either **misogyny** or **misandry**. See also, **racism**.
Sex-linked Sex-linked local groups	The **genes** associated with a person's **sex chromosomes** (giving rise to **male** or **female** characteristics). See also, **heterogametic** (X-Y) and **homogametic** (X-X).
Sexual reproduction	Irrespective of the apparatus involved, sexual reproduction amounts to a merging of male and female **genetic** material. See also, **meiosis**.

Shia Shi'ites	Comprises 10-15% of all **Muslims**, whose teachings derive from the descendants of the prophet Mohammed (many of whom are notionally allied to **Russia**, via the **Russo-Shia bloc**). See also, **Sunni**.
Single integrated economy	Noted for its social progress, regional stability and raised living standards (and for economies of scale and increased output).
Situational attribution (psychology)	Arises when **behaviour** is attributed to a person's situation, rather than their **disposition**. The **Left-wing** is inclined towards situational attribution, perceiving the poor as victims of circumstance. See also, **dispositional attribution**.
Social action Social action crucible Social action perspective/s	A **microsociological** perspective, which has parallels with **existentialism**, in-so-far as autonomy, individualism and free will are seen as shaping the social order – most conspicuously, when **anti-exploitation bias** is evident. See also, **biologically-produced**.
Socialism Socialist/s Socialist principles	**Left-wing** socialism stresses a reduction in material inequality, plus access to education and advancement, coupled with **economic** stimulus and **centralization**. Socialists fear that without equality, self-actualization becomes dependent upon inherited wealth, rather than open-competition and merit. See also, **capitalist**.
Socialization Socialize	According to Anthony Giddens (1938-), **social structures** are maintained through top-down socialization (socialization being a **structure**). However, **agents** are free to ignore, replace or amend those **structures**, using **structuration**, including the manner in which people become socialized, i.e. society has a **duality of structure**. Socialization, in a democratic sense, denotes **capitalist** conditioning infused with **socialism**, leading to fewer inequalities and raised living standards. See also, **gender**.
Socialized family	Beyond the **biologically-produced** nuclear family, lies the **socially-constructed** socialized family (neither of which has a monopoly, as regards **aequiponimity**). See also, **functionally-nucleated family**.
Socializing	See **socialization**.
Social learning (peer-to-peer communication)	According to Professor Alex Pentland (1952-) social learning, or peer-to-peer communication, is what actually changes human behaviour (as opposed to **received learning**). As **social action** spans every **neurotypic sub-classification**, social learning is frequently corrosive, leading to **deviance** and **dumbing-down**. See also, **digital determinism**.
Socially-constructed (sociopolitical superstructure)	Arrived at through symbolic socializing processes, i.e. **semiotics**, the range of **social structures** fabricated by humans is incalculable. See also, **biologically-produced** (genetic substructure).
Social power	Professor Michael Mann (1942-) proposes that **social power** takes two principal forms, i.e. **distributional power** and **collective power**.
Social structure/s	See **sociopolitical superstructures**.
Socioeconomic	See **economic**.
Sociological perspective	The **political perspective** informs the sociological perspective, and vice versa (that is, whether a **microsociological** or **macrosociological** interpretation of society is pertinent).
Sociopolitical superstructure/s	Sociopolitical superstructures (also known as social structures or structures) are **socially-constructed**. They span **semiotics**, **gender** and **emotions** (and include the **functionally-nucleated family**, **global financial system** and **science**). See also, **'x' axis** and **agents**.

243

Sodium inactivation Sodium ion channels (neurology)	Openings in a **neuron**'s plasma membrane permit sodium ions (Na^+) to enter the cell, triggering a **nerve impulse**. Sodium inactivation, in the **inactivation-emission** phase, causes those channels to close.
Solidity Solid/s (dark sciences)	Solidity exists due to **dispersion forces** and **dispersive fields** (making liquids, gases and vapours all technically solids, due to the mutual repulsion of their **electrons**). See also, **corporeal reality**.
Sound Sound argument Sound deduction	**Unsound** individuals are either (a) **rationalists** or (b) **empiricists** (rationalists inferring *a priori* from what they cogitate and empiricists inferring *a posteriori* from what they see). *Ergo*, **unsound induction** comprises (a) or (b), whereas sound deduction comprises (a) + (b), i.e. **hypothetico-deductive reasoning** takes the rationalist's *a priori* **working hypothesis**, and tests it empirically in an *a posteriori* manner. See also, *Homo aequipondium epivevatics*, **epivevatics** and **sound mind**.
Sound mind/s Sound-minded Sound-mindedness Soundness	**Sound** statements are more than **internally valid**, in-so-far as they contain **verifiably true premises**. Such sound-mindedness is comprehensive, unambiguous and precise (unlike **unsound** analysis, which is imprecise, ambiguous and lacking). See also, **first measure of soundness**, **second measure of soundness**, **third measure of soundness** and **fourth measure of soundness**.
Sovereign Sovereignty	Supranationalism involves a transfer of sovereignty to a transnational **power**, whereas **devolution** gifts power to regional assemblies, with sovereignty retained by that doing the devolving.
Space Space-time (dark sciences)	Space is a colloquial term for **spatialgravitism**, and the accompanying **gravitational** and **dispersional** fields – fields which are responsible for **galactic** contraction and **intergalactic** expansion respectively. See also, **laws of topodynamics**.
Spatialgravitism (dark sciences)	**Gravitational** and **dispersional** fields interact as **spatialgravitism**, mirroring the way in which **magnetic** and **electric** fields interact as **electromagnetism**. See also, **laws of topodynamics**.
Specialization (linguistics)	Occurs when a word or phrase becomes more narrowly-defined.
Species (taxonomy)	Species are identifiable on basis of substantive genotypic differences. See also, **binomial nomenclature**.
Spectrum of attraction (SoA)	The **gravitational force** gives rise to a spectrum of attraction, i.e. **strong nuclear force**, **weight**, **gravity** and **gravitation**.
Spectrum of repulsion (SoR)	The **dispersional force** gives rise to a spectrum of repulsion, i.e. **weak nuclear force**, **solidity**, **dispersion** and **dispersive fields**.
Speed of light (c)	**Mass** (m) and the speed of light (c) are closely-related, in-so-far as **light** needs **gravity** to propagate. That is to say, in the absence of gravity there's only **deep-space nucleosynthesis**, converting light energy (E) into a medium supportive of the speed of light.
Stabilization measures (economics)	Stabilization measures generally take the form of **fiscal policies**, **monetary policies** and **flexible immigration**. However, under crisis conditions a government may freeze wages, introduce rationing and/or renegotiate their repayment of debts, etc.

244

Stabilizing Stabilizing fecundity Stabilizing selection	Stabilizing selection governs **climax communities**, due to **differential mortality** favouring established **phenotypes**. Stabilizing fecundity, which favours established **neurotypes**, is the dominant feature of a climax community arising from **coevolution**. See also, **demes**, **random selection** and **progressive selection**.
Stagflation (economics)	Occurs when wages and prices increase during **economic** slowdown and recession. See also, **deflation**, **inflation** and **unemployment**.
State/s (politics)	See **nation states**.
States of matter	See **phases of matter**.
Stellar **nucleosynthesis** (dark sciences)	Stellar nucleosynthesis, which pertains to stars and supernova, is a consequence of the **spectrum of attraction** (just as **deep-space nucleosynthesis** is a consequence of the **spectrum of repulsion**). See also, **dark cycle theory of everything** and **black holes**.
Steroidal hormones Sterol/s	Steroid **hormones** are synthesized in the adrenal glands and gonads (from a sterol called **cholesterol**). See also, **arteriotransmission**.
Stratified equity Stratified extrapolation (cruciform calculus)	**Stratified equity** arises due to the fact that accurate **economic** data is more readily available as one ascends through **biased**, **logical**, **sound** and **balanced**. See also, **lateralized equity** and **economic equilibrium**.
Strong nuclear force (dark sciences)	The **gravitational force** conforms to the **inverse square law**, such that the force of attraction between adjacent **protons** is astronomical. **Neutrons** are far more complex, in-so-far as they form from protons and **electrons**, such that they destabilize a **nucleus** through the **weak nuclear force**. See also, **spectrum of attraction**.
Structuralism Structuralist/s Structurally Structural perspective/s	Structural perspectives, e.g. **Marxism**, **neo-Marxism** and **functionalism**, interpret society from a **macrosociological** angle (that is, they see society as being **socially-constructed**). The opposing **social action perspectives**, however, see society as emerging from **microsociological** factors, ones which are **biologically-produced**.
Structuration Structurating	According to sociologist Anthony Giddens (1938–), **agents** influence **structures** through structuration (an impact which can be thought of as **progressive**, **stabilizing** or **deleterious**). Borrowing from Gidden's structuration theory, **cruciform coevolution** comprises structurating agents and **socializing** structures.
Structure-function	See **macro-molecular machinery**.
Structure/s	See **sociopolitical superstructures**.
Sub-aequipine Sub-aequipine behaviour	**Hegemonic democracy** must maintain **cruciform coevolution** if it is to cure itself of **unsound** sub-aequipine elements, given to **pecuniary** and **prestige** extremism. See also, **aequipine**.
Sub-atomic **quadrants**	See **conceptual-QSD**, **conceptual-QED**, **corporeal-QSD** and **corporeal-QED**.
Sub-atomic particles	The principle sub-atomic particles are **protons**, **neutrons** and **electrons**. However, fields such as **supersymmetry** envision many more. See also, **corporealism**.
Subjective Subjective truth	Rationalists infer a priori from what they cogitate, making subjective truth synonymous with **unsound induction**.

sub judice (law)	An allegation shouldn't be publicly debated, lest it prejudice an ongoing judicial case (in the absence of a case being brought, that allegation is simply an aspersion). As that informal debate revolves around **biased** and **logical** inferences, the conclusions publicly arrived at are primarily a reflection on the public. See also, **dumbing-down**.
Subnational (politics)	**Decentralization** empowers subnationally, such subsidiary aiding **social action**. See also, **national** and **supranational**.
Subordinate Subordinate behaviour Subordination	**NEURON** subordinates those who aren't **sound** and **well-balanced**, its **negative-feedback** curtailing unsound **economic**, **political**, **ideological** and **military** powers. Consequently, anything disabling NEURON has the hallmarks of a 'nerve agent', with civilization soon convulsing from accreted **bias** and **logic**.
Sunni	Approximately 85% of Muslims are Sunnis, whose teachings derive from the prophet Mohammed (many of whom are notionally allied to the **USA**, via the **Americano-Sunni bloc**). See also, **Shia**.
Superego (Freudian psychology)	The human brain's conscience to the ego, affecting and influencing **autonomous morality**, **attributional awareness** and **personality**. See also, **id** and **ego**.
Superimposer Superimposition Superimposition of persons	Cognition can be superimposed, provided both party's **absorption profiles** are known (an **unsound** dominant party having the **power** to unethically influence the identity and thought processes of an unknowing **subordinate**). See also, **connubial dualism**.
Superpower status (anachronism)	During the **Cold War** much of the world polarized into pro-American and pro-Soviet factions, reflecting the enormous **power** concentrated by those superpowers. See also, **global governance complex**.
Superstructure	See **sociopolitical superstructure**.
Supersymmetry	A mathematically pragmatic way of resolving orthodox **science**'s inconsistencies, as regards its standard model of the universe (even though weak evidence in support of an unfalsifiable **hypothesis**, causes **science** to move cautiously towards an as-yet-unproven nonsense).
Supply (economics)	Aggregate supply is synonymous with total output. See also, **demand**.
Supply and demand (economics)	The **supply** of goods and services doesn't always keep abreast of **demand**, necessitating **economic interventions**.
Supranational (politics)	Transcending **national** and **subnational** politics, and thus necessitating a transfer of sovereignty to a transnational **power**.
Supraordinate Supraordinate minds Supraordinate states Supraordination	To lessen the destructive impact of **subordinate** thinking, one requires **supraordination**, whereby **sound** and **balanced** states, groups and individuals are rewarded with **neuro-cognitive** advantages. The most commendable **states** forming a **global governance complex**.
Syllogism	An inference founded upon two premises.
Symbolic interactionism	A **microsociological** position, which sees symbolic socializing processes, i.e. **semiotics**, as shaping society. As those **concepts** are **subordinate** or **supraordinate**, they may or may not prove decisive.
Symmetries Symmetry (physics)	German mathematician Emmy Noether (1882-1935) argued that for every **conservation law** there's a symmetry – a symmetry arising whenever a system changes, but aspects of it remain unaltered.

Sympathetic (nervous system)	The **autonomic nervous system**'s sympathetic branch 'excites' (whereas its **parasympathetic** branch 'inhibits').
Sympatho-adrenal stress response (human physiology)	Stress triggers the release of adrenaline, which acts on the **sympathetic** nervous system, increasing one's heart-rate, deepening one's breathing and dilating one's pupils. Simultaneously, cortisol and glucagon act in concert, raising blood glucose levels.
Synaptogenesis	See **exuberant synaptogenesis**.
Synergy	Combined action, whose effects exceed the sum of the respective parts. See also, **life** and **Gaia theory**.
Syntax (semiotics)	Signs, symbols and words, together with the rules governing their usage. See also, **semantics** and **pragmatics**.
Systemic bias (law)	Arises when **unsound** police officers, or members of the judiciary, identify exclusively with the prosecution. Police brutality having its roots in plebian **social learning**. See also, *Homo sapiens*.
T	
tabula rasa (neonatal 'blank slate')	A *'uniquely-blank, person-specific material, imperfectly suited to inscription'* best describes the neonate **brain**.
'tax and spend' (economics)	Synonymous with **centralization** ('*tax and spend*' strategies are commonly associated with the **Left-wing**).
Tax revenue (economics)	Revenue extracted from private individuals and companies by the **state**. See also, **progressive taxation** and **regressive taxation**.
Telekinesis	The manipulation of objects using only the power of one's mind.
Telencephalon Telencephalonic mind (forebrain)	The **forebrain** comprises the telencephalon (**cerebrum**, **limbic system** and **basal ganglia**) and **diencephalon**. One proposes reclassifying the basal ganglia so that it forms part of the diencephalon (that way the forebrain remains the same, but the proportions of the telencephalon and diencephalon alter). The telencephalonic mind would then correspond, as per the **triune brain**, with the neomammalian complex and paleomammalian complex (it being synonymous with the **CN-cycle**, or **mind cycle**).
Telepathically -induced effects Telepathy	The range of effects achievable by means of radiofrequency technologies (namely, **remote manipulation**, **mental telepathy**, **superimposition of persons**, **psychokinesis** and **telekinesis**).
Tensor-poiesis	Using **syllogism**, one can adduce that Albert Einstein (1879-1955) subscribed to **neo-Platonism**, in-so-far as he used tensor calculus to comprehend **space-time** (implying that reality really can be modelled using maths). A tensor being a mathematical 3D-object, or space, which one can explore from different positions, and poiesis meaning 'to create'. Begging the question, can one conjure-up geometrical fields and arithmetical **symmetries**, sufficient to bring algebra to **life**?
Tentative answer/s	A **working hypothesis** which isn't falsified, in spite of rigorous experimental testing, becomes a tentative answer.
Tertiary responsibility	The tertiary responsibility of **well-balanced** individuals is to establishing an equitable **economic** infrastructure, incorporating a judicious **decomposition of capital**.
Theism Theocracy	Belief in a supreme deity or God.

247

Theory of everything	See **dark Cycle theory of everything**.
Thermodynamics	See **laws of thermodynamics**.
Third law of thermodynamics	The third law of thermodynamics states that there's a minimum temperature at which the motion of the particles of matter would cease. The **dark sciences** argue that **electrostatic attraction** increases as **energy** wanes, so maybe there's a minimum temperature at which the motion of charged particles is replaced by stationary **neutrons**?
Third measure of soundness	Appreciating how information is received or comprehended by third parties. See also, **first measure of soundness**, **second measure of soundness** and **fourth measure of soundness**.
Thought/s	More fleeting than **brain waves**.
Top-down (political action)	Seen from a **structural perspective**, society is the product of **ruling class ideology**, tinged with **anti-insurgency bias**.
top-down (cognitive processing)	**Conceptual processing** (producing **efferent signals**, which **rationalists** place supreme weight and importance upon).
Transcription	See **protein synthesis**.
Transculturation	Occurs when a culture is exported, and the local population sidelines its traditional ways. See also, **acculturation** and **socialization**.
Transduces Transduced	The conversion of **energy** from one form into another.
Transfer RNA (tRNA)	See **ribosome**.
Transistor/s (electronics)	Transistors replaced valves in the 1940s (the transistor's base using a small current to manipulate larger flows of **electricity**). In a similar way, **neuronal induction** enables a **neural pathway** to reverberate.
Translation	See **ribosome**.
Transverse wave/s (physics)	A wave which affects the **medium** through which it travels, in a manner which is at right angles to the direction of travel.
Trinomial nomenclature (taxonomy)	Below **genus** and **species** one proposes a categorization based upon **neurotypic sub-classifications** (for example, *Homo sapiens epipolioticus*). See also, **binomial nomenclature**.
Triune brain	See **placental mammals**.
tRNA (transfer RNA)	See **ribosome**.
Trophic level/s	A food chain's trophic levels include producers (plants), primary consumers (herbivores) and secondary consumers (carnivores). **Primary endosymbiosis** postulates that **mitochondria** and **chloroplasts** gifted nucleated life **ATP** (paving the way for larger multicellular life-forms, longer gestation periods, enhanced anabolic reactions and additional trophic levels).
Two-party system (politics)	Whilst the two-party **political** system makes gaining a working majority easier, it also increases the risk of elective dictatorships (ones motivated by **political bias** and **political logic**). Also, a party may enter into coalition with a party which isn't representative.
U	
UK	See **United Kingdom**.
'uncertainty principle'	See **psychostationary standing** and **psychodynamical momentum**.

Unconscious competency	The ability to perform tasks subconsciously, whilst doing other things (that action being imprinted onto one's nervous system, via **efferent signals**). See also, **basal ganglia**.
Underage brides (child brides)	State-sponsored paedophilia (as evinced by '*Islam's child bride photos*' on the internet). Arguably, condemnation of **Israel** should be made contingent upon its cessation. See also, *Homo sapiens*.
Underproduction (economics)	Supply-side perturbations (arising from an energy crisis, climatological misfortune, ill-judged corporate tax burden, 'no-deal' Brexit, etc) could trigger rising **unemployment** and **inflation**.
Unemployment (politics/sociology)	It's a tragic irony, but many jobless claimants who take their own lives are probably highly-conscientious prospective employees. That is to say, being '*welfare system averse*' isn't the same as being '*work shy*'. The only civilized sanction, in the case of joblessness, is a brief period of employment, sufficient to demonstrate availability. See also, **dispositional attribution**.
Unistable multivibrator (neuroscience)	The unistable multivibrator consists of two or more **neurons**, which are able to excite one another in an inexhaustible manner (with adjacent neurons serving as an **axonal input** and **dendrital output**). Provided the axonal input is rendered redundant by **apoptosis**, the multivibrator then exists as a 'bit', or **nubit**, of one's **memory**. One should note that each 'bit' comprises a one or a zero (in other words, the multivibrator may not reverberate at all). To access that 'bit' of memory, the dendrital output must be stimulated with a different frequency, enabling the conscious mind to establish whether a one or a zero has been 'saved'. See also, **multivibrational psychometrics**.
United Kingdom (UK)	Great Britain incorporated Ireland, via the Act of Union (1800). The Anglo-Irish war (1919-21) then created a United kingdom comprising the British mainland and Northern Ireland. In the 1990s, **devolution** provided for assemblies and parliaments in Northern Ireland, Scotland, and Wales. Provided that the Northern Ireland assembly reconvenes, following its cessation, a more representative UK should emerge.
United Nations (UN)	Formed in 1945, the UN is tasked with preserving global peace and security (and with facilitating cooperation between member states).
United States of America (USA)	Fully two-thirds of all US presidents have been **Right-wing** Republicans. However, **globalization** cleverly frustrates those not occupying the **political middle-ground**. That is to say, a stubbornly Right-wing America will find itself prudently frustrated.
Unnatural fiscal selection (economics)	**Balanced-minds**, supported by **sound** data, are able to sustain **economic equilibrium** (unlike **ill-balanced** others).
UN Security Council	See **Security Council**.
UN Security Council Resolution 242	A **United Nations** resolution, adopted in 1967, demanding that Israel withdraw its armed forces from those territories occupied in the Six Day War (a resolution which the Jewish Virtual Library says has been 91% adhered to, due to Israel's withdrawal from the Sinai Peninsula).
Unsound Unsound induction Unsound factions	**Rationalists** infer *a priori* from what they cogitate, whereas **empiricists** infer *a posteriori* from what they see, i.e. both mindsets exhibit unsound induction. See also, **sound deduction**.
US/USA	See **United States of America**.

249

V

Valence electron/s Valence shell/s	Those **electrons** which are located in the outermost shell of an atom, and which are directly involved in chemical bonding.
Valid Valid argument/s Validity	Valid arguments have conclusions which follow **logically** from their premises (which has its corollary in **programming logic** and **political logic**, where the envisioned outcome is logically arrived at).
Values (mathematics)	Mathematical objects used in calculations (such numbers can be real, natural, complex, irrational, imaginary, etc). See also, **logicism**.
Vapidity (dark sciences)	The **dark sciences** define **solidity** as the mutual repulsion of **electrons** (making gases, liquids and vapours all technically solids).
Variable/s (mathematics)	See **dependent variable**, **independent variable** and **mutually dependent**.
Variable/s (logic)	See **formalized logic** and **predicate calculus**.
Verifiably true premises Verificationism	A **sound argument**, by definition, is one which is **internally valid** and contains verifiably true premises. Such verificationism is exemplified by **hypothetico-deductive research** and the **scientific method**.
Vertical axis	**Cruciform calculus** utilizes the **x-y coordinate system**, i.e. a vertical 'y' axis for **neurotype** and a horizontal 'x' axis for **sociopolitical superstructure**. See also, **horizontal axis**.
Viruses	Exceptionally small non-living **genetic** parasites, consisting solely of **DNA** and **RNA** surrounded by a **protein** coat, which are dependent upon a host cell's **organelles** for survival and reproduction. Cutting-edge gene therapy is currently trialing custom-made viruses.
Voltages	**Boolean operators** process **binary inputs**, i.e. low and high voltages.

W

Wage-price cycle (economics)	People anticipate **inflation**, not **deflation** (hence the demand for higher pay, and with it a wage-price cycle). Competing salaries aggravate the inflationary wage-price cycle. Making the '*tax and spend*' solution, revered by **socialists**, a **sound** way to fund **NEURON**.
Wall Street Crash (1929)	New York's Stock Exchange crash, in 1929, confirmed America's global economic importance (not least, because it brought about the collapse of the foremost German banks). **Logically**, to break-free of that nascent **globalization**, Adolf Hitler (1889-1945) needed to subordinate the world economy. See also, **unnatural fiscal selection** and **neurotypic sub-classifications**.
Wave intensity Wavelengths	Wave intensity refers to the amount of **energy** carried by a wave, and is a reflection of its **frequency** and **amplitude** (its wavelength being the distance between two successive crests, expressed in metres).
Wave-particle duality (physics)	The view that particles possess wave-like characteristics – and **electromagnetic radiation**, the characteristics of so-many particles.

250

Weak nuclear force (dark sciences)	Neutrons form from **electrons** and **protons**, when **energy** is absent and **electrostatic attraction** is overwhelming. However, those electrons are linked to a whole **spectrum of repulsion**, due to the **dispersional force**, which increases according to the **direct square law**. *Ergo*, a weak nuclear force exists between adjacent **neutrons**, initiating radioactive decay and affording neutrons a role in **fission**).
Weight	See **spectrum of attraction**.
Well-balanced	Conforming to **the rule** appropriated from childcare, which says that one should "*favour rewarding; using punishment sparingly, and only to define the limits*". See also, **ameroliptics** and **ill-balanced**.
West, The Western Allies Western polyarchies Western science	The West is an abbreviation of Western Allies, i.e. those fighting on the Allied side during World War Two (and who went on to form **NATO** and its affiliates). This book is a response to its strategically-motivated scientific orthodoxies, many of which veil corrosive social learning.
WOMEN IMPART (social campaign)	The writer's **heteronnubial** co-author (who epitomizes the contribution that **ameroliptical** women can make, as regards **remote manipulation**, **mental telepathy** and the **superimposition of persons**) proved instrumental when conceiving of WOMEN IMPART.
Work (physics)	A transfer of **energy** to an object, which has been subject to a force.
Working hypotheses Working hypothesis	Employing a **hypothesis** which isn't strictly falsifiable could leave **science** advancing towards an as-yet-unproven nonsense (as the law commonly uses unfalsifiable hypotheses, weak evidence risks tipping the balance in the direction of irrefutable nonsense).
X	
'x' axis	See **horizontal axis** and **x-y coordinate system**.
x-y coordinate system, The Cartesian	In **cruciform calculus**, the **vertical axis** represents the **biologically-produced** genetic substructure, or **neurotype**, and the **horizontal axis**, the **socially-constructed** sociopolitical superstructure (more commonly termed **agents** and **structures** respectively).
Y	
'y' axis	See **vertical axis** and **x-y coordinate system**.
Z	
'zeroth law' of thermodynamics	States that if two thermodynamic systems are in thermal equilibrium with a third, they must also be in thermal equilibrium with one another.
Zionism (politics/history)	Campaign aimed at establishing a permanent Jewish homeland in British-controlled Palestine (formerly Judea). See also, **Israel**, **anti-Semitism** and **UN Security Council Resolution 242**.
Zygote (embryology)	A zygote forms when male and female **gametes** fuse.

251

NOTES

1   Commensurate with adopting the appellation *'dark sciences'* in this book, a *'key astrophysical question'* was put to the UK Parliament's Science and Technology Committee (scitechcom@parliament.uk, 27 Jul 2016).

2   According to the dark sciences, *'energy'*, not *'distance'*, is the true relationship between 1^+ and 1^- (with *'distance'*, in Coulomb's Law, being an illusory correlation).

3   The proposition that intelligence denotes learning and adaptation comes from Richard L Gregory Ed, *The Oxford Companion to the Mind* (Oxford University Press, 1987), p375-379 *'Intelligence'*.

4   Homeostasis, as depicted in 'Figure 5: Homeostatic control process', draws upon ideas outlined in MBV Roberts MA PHD, *Biology: A Functional Approach,* second edition (Thomas Nelson and Sons Ltd, 1980), p204.

5   Evidently, the terms 'science' and 'scientific' could be used pejoratively to denote less than balanced. In fact, a person calling themselves *'a scientist'* could be biased, logical, sound *or* balanced (given the term's common or colloquial usage).

6   Reasoning lies at the heart of Paul Tomassi's book *Logic* (Routledge, 1999). The perfect book to begin arguing the case for biased, logical, sound and balanced thinking (including logical form, sound arguments, hypothetical reasoning, etc).

7   In the absence of litigation, the term aspersion is applicable. In other words, it's an aspersion until it's litigated, whereupon certain legal protocols apply (not least of which, *sub judice* restrictions).

8   Notwithstanding the immense scientific value of mathematics, it's impossible to represent one third of 10 using the decimal system, i.e. 3.33r x 3 = 9.99r (a mathematical proof that reality resists arithmetical expression). In fact, mathematics contradicts the fundamental laws of physics, in-so-far as there's a law of conservation of mass which states that *"matter can neither be created nor destroyed"* – and yet, if one takes a 10m strip and cuts it cleanly into three *equal* lengths, the sum of those arithmetically-equal parts is less than 10m. That is to say, anything up to 0.01m of matter has simply vanished or otherwise been destroyed. Such that if one accepts this clinical refutation, then orthodox science needs to urgently revise its present relationship with algebra.

9   Employing the model proposed by one's heteronnubial co-author – namely, *"that bar magnets influence one another's magnetic fields, such that electromagnetic induction produces variable electromotive forces on their respective electrons, sufficient for them to attract or repel"* – the writer suggests that covalent bonding follows much the same pattern, with atoms drawn together or otherwise repelled.

10 Biasing and debiasing, plus declarative and procedural knowledge, are discussed in Jonathan St.BT Evans, *Bias in Human Reasoning: Causes and consequences* (Lawrence Erlbaum Associates, 1994), p66-70 *'Knowledge and reasoning'*, p70-79 *'Biasing effects of Knowledge'*, and p79-86 *'Debiasing effects of knowledge'*.

11 Regrettably, websites seldom specify that they are legally compliant. Such that the internet may be becoming more illicit through time, with only limited information by which to judge (comment submitted to Labour MP, Cat Smith, in 2018).

12   In essence, Left-wing incitement stems from *situational attribution*, e.g. the NUM and Tony Benn MP (1925-2014) campaigned for the release of two mineworkers convicted of murdering taxi driver David Wilkie (1949-1984). Although later reduced to manslaughter, it illustrates the Left-wing's tendency to exonerate individual's 'murdering' on account of hardship. Of course, the original conviction for murder may have been Right-wing scapegoating, arising out of *dispositional attribution*.

13   Sociological power is a complex subject, which can be approached from many different angles, e.g. legitimate power, expert power, informational power, reward power, coercive power, etc. The angle adopted in this book derives from Michael Haralambos's and Martin Holborn's, *Sociology: Themes and Perspectives* (Collins Educational, 1995), p544-546 *"Michael Mann – The Sources of Social Power"*.

14   The 'Future Separation of Power' illustration, in Mark Fox's, *Destructive Interference: Understanding the brain's telepathic potential* (AS-Publishing, 2014), p.93, makes reference to the 'political' (i.e. political, economic and ideological powers).

15   On the 27 June 2016, the writer, with their co-author's approval, submitted a '*second EU referendum request*' to the then Prime Minister, David Cameron MP (and the opposition leader, Jeremy Corbyn MP), worded as follows: *"The EU referendum result (a result which could not have been accurately predicted) has generated a raft of significant national and international questions, any of which might lend themselves to major election pledges and/or further referenda. For example, we are now faced with the prospect of those pernicious extremist views, responsible for the late Jo Cox's murder, being actively rewarded; the very real possibility of the IRA realizing its stated war aims; the wholesale loss of Gibraltar, due to the strength of that particular vote; and, during the centenary of the First World War, the potential for the UK and EU to fragment, in the face of resurgent German nationalism. Given the profundity of these concerns, we mustn't merely assume the thoughts, feelings and opinions of the electorate. Some have argued that we should simply respect the first referendum result; but this implies that there is no scope for revising that result, in the light of what we now know."* Central to that submission, was the question of whether a judgement exists in perpetuity. Or whether, as seems reasonable, one can revise such decisions in the light of new information.

16   Were 20th century global conflicts and the earlier Seven Years War, at root, European wars, as professed by Niall Ferguson, *Empire: How Britain Made the Modern World* (Penguin Books, 2003), p 34; or were they, as Constructive Interference argues, wars rooted in imperialism, begging its antithesis?

17   Michael Balfour, *Germany: The Tides of Power* (Routledge, 1992), p115. Describes how a once rapaciously *competing* nation became extraordinarily *competitive*.

18   UK independence was muted in the previous book – namely, Mark Fox, *Destructive Interference: Understanding the brain's telepathic potential* (AS-Publishing, 2014) – but that stance was predicated upon the United Kingdom's anticipated role within NATO (and the status of dollars, pounds and euros worldwide). As things stand, logical and biased reasoning, by ill-informed Brexiteers, may well prove ruinous.

19   In 2018, the Nobel Peace Prize was awarded to Nadia Murad (1993-) and Denis Mukwege (1955-), in recognition of their efforts at ending the use of sexual violence as a weapon of war. However, partisan reporting made a self-interrogation of one's own armed forces appear unnecessary (ironic, given the scandal of US Marines sharing illicit photos of female colleagues and British servicemen having raped and murdered Louise Jensen (1971-94), a Danish tour guide, in 1994). In other words, an unconscious bias was banefully present in the media's reporting.

20   William J Baumol and Alan S Blinder, *Economics: Principles and Policy* fourth edition (Harcourt Brace Jovanovich, 1988), p81-83, Figure 5-4 '*The Inflation Rate in the United States, 1870-1987*', p209. Deflation all but ends after World War II, which some attribute to minimum wage controls, unionization and interventionism (*ergo*, sound and balanced interventions aren't ineffectual, and could be developed).

21   The Nuremberg Trials (1945-49) established that perpetrators can't hide behind the assertion that they were only obeying orders, i.e. it would be wrong to chastise Tony Blair (1953-) and not reprimand those personnel who chose to comply.

22   Different neurotypic sub-classifications interpret cruciform coercion in different ways, and one should avoid unsound interpretations (the balanced-minded favour rewarding outcomes, where necessary building upon biased and logical activity).

23   Stella Cottrell, *Critical Thinking Skills* (Palgrave Macmillan, 2005). Comprehensive introduction to argument, proof and reason.

24   It's not difficult to find *conspicuous consumers*, only *conspicuous connoisseurs*.

25   Resistance needn't target people, it can target the infrastructure of competing regimes, e.g. oil refineries, railroad tracks, bridges, etc. Which some would argue is a sparing use of punishment, when faced with pecuniary or prestige extremism (activities which the vegetarian, T.E. Lawrence (1888-1935), would've approved).

26   Gender convergence and divergence are macrosociological explications, which pertain solely to top-down ruling class ideology and ascriptive practices (they do not pertain to, nor seek to impose upon, individual choice at the microsociological level).

27   https://www.unicef.org/newsline/00pr17.htm (UNICEF Executive Director Carol Bellamy, in the run-up to International Women's Day, 07-Mar-2000, stated "*According to 1999 estimates, more than two-thirds of all murders in Gaza strip and West bank were most likely 'honour' killings*").

28   One's heteronnubial co-author has perceptively contributed to Constructive Interference. For example, hypothesizing that the confluence of two bar magnet's electric and magnetic fields produces an electromotive force on their respective electrons, sufficient for them to attract or repel; conjecturing that electromagnetism's spectral progression through time made life conceivable; arguing that the biologically-produced subverts the socially-constructed, unless that which is socially-constructed subordinates *Homo sapiens*; and suggesting that the manner in which a potential difference arises in electronics may have relevance in biological systems. An 'opportunity' is a situation which lends itself to being rewarding, hence one's recognition of having matured creatively under their balanced influence (what's more, in the newly devised field of ameroliptics, they're an unquestioned authority).

29   The law, by not behaving in a '*straight-up-and-down*' manner, weakens the case for contrition. Moreover, by acting dishonourably, it guarantees that people will be more vexatious, not less. So pervasive is social learning, that we find individuals working in today's courts, police and law who'd readily alter sworn depositions. Of course, many of them would argue that their actions made little material difference (notwithstanding, that the law wouldn't concede to drivers exceeding the permissible speed limits on similar grounds). One must conclude that the law is what happens when one places sound and balanced legal textbooks in the hands of *Homo sapiens*.

30   Liberal Democrat MP, Norman Lamb, who's presently chair of the Science and Technology Committee, has described learning disability and autism care in Great Britain as 'shameful', given its heavy-handed use of face-down restraint, locked seclusion and forced medication. Clearly, more could be done to determine the neurotypic sub-classifications of those employed in that sector.

31   The notion of modelling events at an accelerated pace comes from the Horizon programme *Time Trip* (BBC, 2003); with this book applying that idea to cyclical phenomena like the human mind and cruciform coevolution.

32   In many ways, Constructive Interference is a strongly-worded repudiation of Alex Pentland's, *Social Physics: How Social Networks Can Make Us Smarter* (Penguin Books, 2015), in-so-far as it sees 'Social Physics' as alarmingly optimistic about peer-to-peer communication, both in terms of social outcomes and in terms of quantitative predictive modelling. In truth, cognitive bias and unsound logic plague group psychology, bringing into question the whole notion of "*collective intelligence*". In fact, the Nazi elite would've relished the idea that their shared intellect was greater than they themselves imagined. To summarize, if you take an unsound person and keep adding logical and biased types, you do not – I repeat, do not – arrive at an intelligence which is "*greater than the individual intelligence*".

33   Hamish Barbour Executive Producer, Jutland: Clash of the Dreadnoughts (Timeline World History Documentaries, 2017). Discusses the subversion of life-preserving protocols within British naval circles, with specific reference to cordite handling. The personalities in question aren't unlike those who discard the highway code, on the basis that it could never match their knowledge, skills or experience (in other words, it's what dumb drivers do when they want to appear 'clever').

34   There are intangibles which can't be taught, notwithstanding that teaching can corrupt the most intangible of things. For the record, this book has intangible origins, having been commenced at the behest of one's heteronnubial co-author, who's brought both style and substance to its telling. Their presence proving – not simply that it's possible to experience another's heartbeat – but also that a person could be more generous, but never too generous, in some cases.

35   Lexi Krock, The Final Eight Minutes (www.pbs.org/wgbh/nova/article/final-eight-minutes, 17 Oct 2006). A website containing transcripts of the cockpit voice recordings and a link to the official Spanish accident report, in respect of the Tenerife Air Disaster.

36   The brain's capacity for positively reinforcing beliefs is unlimited, e.g. in aeronautics spatial disorientation is aggravated by deceptive G-forces and the pilot's neurotypic sub-classification. In principle, having eliminated the pilot's situation as the primary cause of events, all that remains is disposition, i.e. the proverbial 'right stuff'.

37 The expressions *'science'* and *'scientific'* aren't strictly scientific, in-so-far as they are commonplace phrases, used colloquially. In other words, a person describing themselves as *'a scientist'* undermines people's confidence in their use of technical language. Fortunately, the terms epipoliotical, hermeneutical, epivevatical and ameroliptical restore clinical precision (enabling so-called 'scientists' to profess epivevatical characteristics).

38 Department for Constitutional Affairs, *Meeting the needs of Customers with Disabilities* (Disability Discrimination Acts 1995 & 2004 – A Guide). Contrasts the medical model and social model of disability, with the latter seeing disability as a form of exclusion from normal life and the former viewing it as a clinical deficiency.

39 This book classifies the hypothalamus as being part of the diencephalon, as per *The Britannica Guide to the Brain* (Robinson, 2008), p.17. That said, different sources routinely define the limbic system, basal ganglia and diencephalon in different ways.

40 Sten Grillner, Brita Robertson and Marcus Stephenson-Jones, *The evolutionary origin of the vertebrate basal ganglia and its role in action selection* (www.ncbi.nlm.nih.gov/pmc/articles/PMC3853485, 14 Jan 2013). Building on the above article's conclusions, the NE-cycle (of which the basal ganglia is part) has conserved features with very ancient origins, whereas the CN-cycle (of which the limbic system is part) has evolved beyond all recognition.

41 MBV Roberts MA PHD, *Biology: A Functional Approach,* Second Edition (Thomas Nelson and Sons Ltd, 1980), p461 (*'Mendel in Retrospect'* explains why Gregor Mendel's work reveals the mind of a genius).

42 Significantly, a person's sex (that is, whether one is male or female) is determined by the presence or absence of a Y chromosome. The Y chromosome being, to all intent and purposes, 'dominant', in-so-far as it eclipses many of those 'recessive' features found on the X chromosome.

43 J Simpkins and JI Williams, *Advanced Human Biology* (Unwin Hyman, 1988), p109-115 (a well-illustrated summary of the cell cycle, which states "*skin cells divide on average once a week, whereas brain cells never multiply*").

44 Contrary to exaggerated claims in support of adult neurogenesis, motor neurons have axons up to 1 metre in length, making binary fission unlikely, if not impossible.

45 Lynn B Jorde and Stephen P Wooding, *Genetic variation, classification and 'race'* (www.nature.com/articles/ng1435, 26 Oct 2004). This article concludes that genetic boundaries are too imprecise to give credence to a genetic concept of race.

46 One has *two* parents, *four* grandparents, *eight* great-grandparents, etc (arriving, after 30 generations or 900 years, at 1 billion). *Ergo*, allele frequencies, in the case of humans, are heavily influenced by the geopolitical infrastructure.

47 The triune brain mirrors the Freudian mind, i.e. reptilian complex (id), paleomammalian complex (ego) and neomammalian complex (superego). On the issue of dreams, environmental factors may lead to a life-preserving disruption of one's sleep, due to dark thoughts playing on one's mind.

48 Axially-advantaged connubial dualists are attracted by enormous plasticity.

49 Answer: both.

50   This book is published in the **dark sciences**' centenary year, as 1919 witnessed the culmination of Ernest Rutherford's so-called *'gold foil experiments'*, whereby alpha particles (comprising 2 protons and 2 neutrons) were fired at gold leaf, whereupon approximately 0.01% bounced back. As gold nuclei contain 118 neutrons, that experiment was really about the mutual repulsion of the neutrons (as each of those neutrons comprised, or otherwise formed from, an electron and a proton). That is to say, neutrons contribute to a spectrum of repulsion, arising from dispersional fields (fields centred upon mutually-repelling electrons). In close proximity, multiple neutrons experience a weak nuclear force, destabilizing very large nuclei – those 'weak nuclear forces' increasing markedly in the case of approaching alpha particles (those alpha particles simply not being close enough for the 'weak nuclear force' to be truly weak). Therefore, Rutherford had – according to the dark sciences – experimentally furbished *strong* evidence for a dispersional field, but only *weak* evidence for positive charges repelling.

51   As coding and electronics both rely on an unassuming logic, teaching them together might avoid railroading children into computing.

52   Sociology radiates overlapping interpretations. Somewhat audaciously, this book tidies-up the subject, painting *positivists* and *amerolipticists* as optimistic and *anti-positivists* and *phenomenologists* as pessimistic, as regards sociology's long-term scientific credentials (the former perceiving aequipine structures as aiding predictive modelling and the latter perceiving sapient agents as frustrating the same). And so, whilst amerolipticism equates with the coevolution of accurate predictive modelling, in areas such as anthropobionomics, phenomenology warns of an increasingly unpredictable future, fraught with escalating ecological uncertainty.

53   The subconscious impact of radiological abuse is difficult to ascertain, further complicating the issue of attribution. All the more reason, therefore, to uphold balanced protocols regarding the use of telepathically-induced effects.

54   In a letter from the Principal Registry of the Family Division, dated 4 Feb 1997, the writer was authorized to *"administer any oath or take any affidavit required for the purpose of proceedings in the Supreme Court by virtue of Order 32 rule 8 of the Rules of the Supreme Court, in England and Wales"*. Such was the pace of restructuring within the English Courts at that time that the writer can rightfully claim to have worked for the Lord Chancellor's Department, Principal Registry of the Family Division, Department for Constitutional Affairs, Supreme Court Group, Courts Service and Ministry of Justice. And, whilst this book may provoke outrage within those hallowed institutions, it's a much-needed wrangle in the eyes of those experiencing unsound levels of remote interference.

55   A letter dated 10 Jan 1998, which was penned by the writer and which précised IMPART's concerns, was sent to the Department of Health (who responded on 3 Feb 1998). On 11 Jan 2008, IMPART forwarded a letter to the then Prime Minister, Gordon Brown MP (who responded on 22 Jan 2008). Also, on 1 Jun 2011, a WOMEN IMPART leaflet was posted to the editor of The Times. Additionally, letters, leaflets and copies of the previous book, *Destructive Interference: Understanding the brain's telepathic potential*, have been sent to prominent institutions.